Chasing Moore's Law

Information Technology Policy in the United States

Edited by William Aspray

Published by SciTech Publishing Inc.
Raleigh, NC
www.scitechpub.com

Printed in the U.S.A.
10 9 8 7 6 5 4 3 2 1

ISBN 1-891121-33-2

SciTech President: Dudley R. Kay
Production Services and Composition: TIPS Technical Publishing, Inc.
Copy Editing: Cavanaugh Editorial Services
Cover Design: WT Design

This book is available at special quantity discounts to use as premiums and sales promotions, or for use in corporate training programs. For more information and quotes, please contact:

Director of Special Sales
SciTech Publishing, Inc.
7474 Creedmoor Rd. #192
Raleigh, NC 27613
Phone: (919) 866-1501
Email: sales@scitechpub.com
http://www.scitechpub.com

Contents

Author Biographies

William Aspray is Rudy Professor of Informatics at Indiana University in Bloomington, where he studies the historical, political, social, and economic aspects of information technology. He holds bachelor's and master's degrees in mathematics from Wesleyan University and a doctorate in history of science from University of Wisconsin–Madison. He previously taught at Williams, Harvard, Penn, Rutgers, Minnesota, and Virginia Tech. He has served as a senior administrator at the Charles Babbage Institute for the History of Information Processing, the IEEE Center for the History of Electrical Engineering, and the Computing Research Association.

David Bruggeman is the Research Associate for the Forum on Information Technology and Research Universities at the National Academies in Washington, D.C.. The Forum is part of the Academies' Government-University-Industry Research Roundtable and seeks to foster dialogue both nationally and on campuses regarding the changes brought about by information technology. He is a Ph.D. student in the Science and Technology Studies program at Virginia Tech's Northern Virginia Center. He holds a bachelor's degree in Politics from Whitman College and a master's degree in Science, Technology and Public Policy from the George Washington University. He previously worked for the Academies' Committee on Science, Engineering and Public Policy (COSEPUP), and for NOAA's National Environmental Satellite Data and Information Service.

Eric Fisher is an auditor with the Department of Defense Office of the Inspector General in Arlington, Virginia. He conducts audits and reviews audit policy guidance that impact the program functioning and oversight responsibility of the U.S. military departments. He holds bachelor's and master's degrees in accounting from Virginia Tech and a Master of Public Administration from George Mason University.

Peter Harsha is Director of Government Affairs for the Computing Research Association (CRA), a Washington, D.C.-based organization of 200 North American Ph.D.-granting computer science departments, industrial research labs, and affiliated computing societies. He serves as the computing research community's representative in Washington on matters of federal IT research and development funding and federal research policy. Prior to joining CRA, he served as a member of the professional staff of the House Science Committee, where he tracked federal IT R&D funding and research policy issues for the Subcommittee on Research. He holds a bachelor's degree in English from Hillsdale College.

Steve Mosier is a Program Manager and Technical Fellow with Northrop Grumman Information Technology in Reston, Virginia. He is a Ph.D. student in the Science and Technology Studies program at Virginia Tech's Northern Virginia Center. He holds a bachelor's degree in Electrical Engineering from Case Western Reserve University, and master's (Electrical Engineering) and MBA degrees from Penn State. He is a member of the Institute of Electronics Engineers (IEEE) and the International Council on Systems Engineering (INCOSE). Steve was among the co-authors of the EIA/IS 731 Systems Engineering Capability Model standard. His research interests include policy and practice of technology transfer and technology diffusion.

Jolene Kay Jesse is a Senior Program Associate in the American Association for the Advancement of Science's Directorate for Education and Human Resources Programs where she conducts research on science, technology, engineering, and mathematics education and policy issues, with special emphasis on the representation of women and underrepresented minorities in those fields. She received her Ph.D. in political science from the University of Wisconsin–Milwaukee, and a master's from The American University in Washington, D.C.

Lorraine Woellert is Legal Affairs Correspondent at *Business Week*, where she covers Congress. In her 15 years as a journalist, she has reported on policy debates relating to trade, health care, intellectual property, and aviation. She has covered two presidential elections and broken numerous stories on campaign funding. She is an elected member of the Standing Committee of Correspondents in Congress, the liaison between Congress and the press. Woellert holds a bachelor's degree from George Mason University and is pursuing her master's in Science and Technology Policy at Virginia Tech.

Najma Yousefi is pursuing his doctorate in Science and Technology Studies at Virginia Tech. He earned his bachelor's degree in philosophy from National University of Iran and his master's degree in Economic Policy Management from Columbia University. Representing the Iranian Ministry of Economic Affairs and Finance, he worked with the World Bank on social sector's projects and studies in Iran. Najma Yousefi is currently the Director of Marketing and Sourcing at ISmart, an IT solutions provider serving state and local governments and educational institutions in the mid-Atlantic region.

Preface

William Aspray

Moore's Law states that the number of electronic switches that can be placed on a computer chip doubles every 18 months. It is not a law of nature, like Newton's laws of gravitation, but rather a way to quantitatively describe the pace of innovation in the semiconductor industry. Industry planners believe that this doubling will continue at the same pace; thus they have extrapolated from Moore's Law to plan what their product performance should be one or five years hence. In doing so, they have made Moore's Law a self-fulfilling prophecy. Most importantly for this book, however, Moore's Law stands as a metaphor for the rapid, incessant course of technological innovation that is occurring in the computing and communications field. Humans have difficulty accommodating to such rapid technological change. One of the major themes of this book, and hence the reason for this title, is that humanly created laws and political solutions also have trouble keeping pace with this rapid technological change.

The purpose of this book is to provide an introduction to the main topics concerning information technology policy in the United States. The book begins with a political study of federal funding for research and development. This chapter sets the stage for most of the later chapters in the book. Much of the research that served as a foundation for the Internet revolution of the 1990s, for example, was supported by the federal government.

The computer has been reconceptualized over the years, first as a mathematical machine for scientific and engineering applications, then as a data-processing machine to drive commerce, and most recently as a communication machine. The early conceptualizations were never entirely lost, but they lessen in popularity as new conceptualization occurs. The next three chapters of this book examine this most recent conceptualization. Chapter 2 covers telecommunications policy as it relates to computing, while Chapters 3 and 4 survey policy issues related to Internet governance and Internet use.

The middle section of the book covers three topics that have been of policy concern for many years but have taken on new importance recently. Computer security was once mainly the province of systems administrators, but now that some of our fundamental infrastructures, such as the power and telephone grids, are dependent on the security of computer networks, computer security has become a national priority in the aftermath of September 11. Data privacy has been an issue of public concern since the 1960s, but the stakes have risen with the rise of the Internet and online information sources. Intellectual property issues were once only a minor consideration in the computer field. Patents were not of critical business importance in the computing field; companies stayed ahead instead by their pace of innovation. But since the Internet's inception, business practice patents, trademark infringements through cybersquatting, and copyright infringements through peer-to-peer file-sharing networks all have become important policy issues.

The final section of the book covers three additional topics that have received significant policy attention in recent years. IBM exerted a commanding presence in the computer industry from the 1950s into the 1980s, but as microcomputers and the Internet emerged, the dominant position of IBM eased and pressures for antitrust action eased with them, allowing Microsoft to become the IBM of the 1990s. Chapter 8 traces antitrust actions in the computer industry, with particular attention to the Microsoft case. The second of these policy issues concerns equal access. The Internet provides a new avenue onto the world, but only if you have access. People who are poor, are members of minority groups, or live in rural areas have had significantly less access than the general public. Chapter 9 traces attempts at political solutions to this Digital Divide. The final policy issue concerns the workforce. In the late 1990s that great economic engine known as the dot-com boom was threatened by inadequate numbers of trained workers. Chapter 10 looks at workforce issues and the various legislation related to training and immigration, especially in light of the dot-com boom and subsequent crash.

There are some topics not covered in this book. There is a rich body of telecommunications policy, such as spectrum allocation and universal service. This book only touches upon this policy in passing, covering it only as it relates directly to computing. Especially in the 1980s and early 1990s, there was great concern over export restrictions for supercomputers, encryption software, and other technologies that could be used for military purposes. Because these issues have died down somewhat in recent years and because there is nothing noteworthy about computer exports compared with other kinds of exports that have potential military uses, such as fissionable material or weapons, this topic is excluded. Our focus is primarily at the federal level, and we have covered only in passing political topics that are dealt with primarily at the state and local levels, such as technology-driven economic development projects and shrink-wrap licenses for software.

One of the problems with writing any book about information technology policy is the rapid pace of change in both the technology and the policy. By the time a book has been reviewed, revised, copy edited, printed, bound, and distributed, it is beginning to be out of date. We have addressed this issue in the way we approach our chapters. Every chapter includes timeless material about the history of the policy issue, the players typically involved, and the positions they typically adopt. To the extent possible, we use recent events to illustrate more general themes, rather than treat these events as the main topics in and of themselves.

The average congressional staff member is young, probably still in her twenties. These people do not have the experience that lets them know how similar political problems have been resolved in the past; and they do not have the luxury of time to become subject experts. Thus one intended audience for this book is those who are in the trenches in Washington, working their sixty-hour weeks to create sound policy, and who might find such a compendium handy as background. A second target audience is the computing research community. Having worked from Washington for a number of years, I know that this community is interested but not particularly familiar with the policies that govern their technological creations. This is a second target audience for this book. Finally, we believe that the book is appropriate for use in the classroom. We have focused on practice rather than theory in our chapters, so the book may not be the first choice for use in a political science course. However, this very emphasis on practice will make it more suitable for use in classes in the newly emerging information schools (where I now teach), with their focus on the study of information technology in application and context. The book is also suitable for use in science and technology classes and courses on contemporary culture.

This book originated from a seminar on information technology policy I taught in the spring semester of 2002 at Virginia Tech's graduate school in suburban Washington, D.C. I have taught at a number of Ivy League and major public research universities, but this seminar was one of the highlights of my teaching career. The students were almost all working adults, bringing both maturity and experience to the course. The majority of them had day jobs in either the computing or policy fields, and they brought material from their work lives into the classroom. For example, the Microsoft antitrust case was under way during that semester, and one of the students whose employment involved tracking the trial would report at each class meeting on what had happened in the trial that week. I invited some of these students to contribute chapters and rounded out the volume with chapters from acquaintances in the Washington policy community. While some of these contributors have not yet made their national reputations in the policy area, I expect that many of them will do so. They are young and bright, and they worked hard to adopt a common style across their chapters, making this a unified book and not simply a collection of essays. It was enjoyable and exciting to work with them.

Chapter 1

IT Research and Development Funding

Peter Harsha

"Where there is no vision, the people perish."

—Proverbs 29:18, inscribed in the hearing room for the
U.S. House of Representatives Committee on Science

On February 4, 2002, President George W. Bush unveiled his fiscal year 2003 budget proposal, a plan for spending nearly $2 trillion in federal funding packaged in a flag-bedecked stack of documents over five inches tall. The stack—a compendium of justifications and accounting for every dollar of spending at every federal agency and office—represented the culmination of nearly a year of work by the executive branch and its agencies, but only the beginning of the annual cycle of debate and legislation necessary to fund the operations of government.

The president chose to mark the release of his budget with an appearance in front of a cheering crowd of fatigue-clad U.S. Air Force personnel, gathered in a large empty hangar at Eglin Air Force Base in Florida. With its release coming just six months after the terror attacks of September 11th and not long after the military successes in Afghanistan that followed, the President's budget request had to be quickly reworked to reflect the new priorities of the day. "The budget for 2003 is much more than a tabulation of numbers," a notice accompanying the release said. "It is a plan to fight a war we did not seek—but a war we are determined to win."

In front of this appreciative crowd, many of whom had recently returned from duty in Afghanistan, Bush announced the centerpiece of his budget plan—a $48 billion increase in military spending, the largest such increase in twenty years. Over the applause of the service personnel, Bush had a message for Congress that would soon take up his budget plan, "We're unified in Washington on winning this war. One way to express our unity is … to set the military budget, the defense of the United States, as the number one priority and fully fund my request."

On the same day in Washington, the President's Office of Management and Budget (OMB), was supervising the distribution of copies of the President's budget request to members of Congress, congressional staff, and leadership offices on Capitol Hill, where the budget request faced an uncertain future. Though the President is statutorily required to submit a budget to Congress by the first Monday in February every year, Congress is under no obligation to consider it.

This year, however, the President's call for increased military spending in fiscal year (FY) 2003 was not likely to be opposed by the majority of those in Congress. With the September 11th attacks still fresh in the nation's memory, the ongoing operations in Afghanistan, and possible future deployments in Iraq, the Philippines, Yemen, or elsewhere in the war on terrorism, Congress was favorably disposed to provide broad support for military funding.

However, the practical effect of such a large increase in military spending was less generous support for spending in a broad range of other accounts within the budget request. Among those who felt slighted by this change of priorities were members of what author Daniel S. Greenberg has called "The Metropolis of Science," whose members include not only the traditional participants in the scientific enterprise—university, federal, and industrial researchers—but also the growing infrastructure that supports them, including federal, state, and local science and technology agencies, university technology transfer officials, business offices, and others.

Though the president's FY 2003 budget plan contained the largest ever request for federal research and development funding—over $112 billion total, an increase of $8.9 billion or 8.6 percent over FY 2002—the scientific community was less than enthusiastic about the proposal. In their analysis of the president's budget request, the American Association for the Advancement of Science (AAAS) noted:

[T]he proposed increases for the Department of Defense (DOD) ($5.2 billion) and National Institutes of Health (NIH) ($3.7 billion) would make up the entire $8.9 billion increase, leaving all other R&D funding agencies combined with barely the same amount as FY 2002. Thus, this would further reinforce the "missiles and medicine" profile that federal R&D has assumed in recent years. Unlike last year, when most of the other R&D funding agencies would have seen their R&D funding decline, FY 2003 would see a mix of increases or decreases averaging to zero growth.

Included in the president's research ad development (R&D) proposal was his request for funding for information technology R&D—a $1.9 billion request shared by 11 different agencies and organized as the Networking and Information Technology Research and Development program (NITRD). NITRD is the largest of the federal government's interagency research and development initiatives, and one frequently cited by members of Congress as representative of the appropriate role the federal government should play in supporting research. However, the President's request called for only a $46 million increase in NITRD funding for FY 2002—a 2.5 percent increase overall. Within NITRD, only the National Aeronautics and Space Administration (NASA) and the Department of Health and Human Services (HHS)—primarily the National Institutes of Health (NIH)—received requests for increases greater than 0.3 percent (NASA receiving $32 million or 17.7 percent more in FY 2003, and HHS receiving $26 million or 8.4 percent). Taking into account an annual inflation rate of about 2 percent, critics noted that the president's request was in reality only a marginal increase overall, and could be construed as a cut to almost all of the participating agencies.

And so the stage was set. The release of the president's budget request that first Monday in February set in motion a series of procedural activities that would focus the attention of the Congress and the Administration on completing a budget agreement and appropriating funding for FY 2003, which would begin October 1, 2002. It also began the debate among stakeholders in the budget, the various constituencies nationwide (and worldwide) who hoped to partake of federal support, over the proper distribution of the $767 billion requested in discretionary spending. (The balance of the federal budget is non-discretionary, spending that is statutorily obligated for such things as interest on the national debt, and Social Security and Medicare payments.) In this context, the constituency of science finds itself in competition with many other constituencies for a share of the discretionary budget: veterans' groups, groups representing senior citizens, advocates for the homeless, consumer groups, groups promoting tax relief, and environmental groups, to name a few. Throughout 2002, the scientific community employed a number of arguments—nearly all well-tested through 50 years of federal science funding (science in the service of national defense, as a protector of health and welfare, and a driver of U.S. industry) and some novel (science as necessary for the president's tax cut—to make its case for priority in the budget. This chapter will examine the development of those arguments over the last half-century of federal support for science in the process of examining the growth of federal support for science over the same period.

Within the science constituency, there was and is competition for priority as well. Congressional and administration policy makers, in an effort to provide some justification for splitting the "R&D pie" in different ways, have helped foster divisions between elements of the sciences—between the life sciences and the physical sciences, between basic researchers and applied researchers, between civilian researchers and defense-related researchers, and a host of smaller divisions. The growth of computing research—from small grants for mathematics research to the largest R&D initiative in the federal budget—can only be understood by examining the interplay of forces acting on the discipline from within and without. This chapter will examine how external factors such as the Cold War and the launch of Sputnik through the attacks of September 11th have shaped the funding, and hence the discipline, through the years. It will also examine some of the forces from within, the people who helped drive the computing research agenda that, in turn, helped create the programs that exist today.

World War II and Federal Support for Science

The conventional wisdom in the aftermath of the 2001 conflict in Afghanistan is that the superior technology of the United States and its allies enabled the quick dispatching of the Taliban regime just 10 days after the start of combat operations. A relatively small contingent of American Special Forces on the ground armed with laser target designators and satellite communications were able to order thousands of tons of explosive ordnance, guided by a constellation of satellites in orbit, delivered by aircraft based thousands of miles away, precisely on Taliban targets. The Taliban, which had held the advancing armies of the Soviet Union at bay for more than a decade until the Soviet pullout in 1989, succumbed in just over a week to the highly accurate bombardment and the advance of local fighters allied with the United States. The U.S. victory, it is assumed, was owed in major part to the significant advantage it possessed in information technology—communications, global positioning, intelligence, advanced electronics, and the ability to marry all of these technologies to obtain a full, accurate picture of the "battle space" and act on it. Indeed, this perception has been seized upon by many in the "metropolis" of science to advocate for continued strong support of the sciences, especially basic research.

Prior to World War II, basic research in areas unrelated to agriculture, geological science, or the aeronautical sciences was primarily funded by private philanthropy and concentrated in a handful of universities and private research institutes.

During the war, these civilian research efforts were put to use by the Roosevelt Administration, which coordinated their work through the newly created Office of Scientific Research and Development. The office led researchers in efforts ranging from atomic bomb research, to work on radar, proximity fuses, and the development of certain antibiotics. Vannevar Bush, the physicist who ran the office (and who lobbied for its creation) believed these wartime innovations argued for the importance of the federal government in continuing to support basic research activities in order to preserve the nation's military and economic strength in the post-war years. In a report commissioned by Roosevelt in November 1944, Vannevar Bush laid out plans for the creation of a new, federally funded but independently run "National Research Foundation." The 1945 report, Science, The Endless Frontier, proposed an agency that would focus on basic research as distinct from "applied" research, and that would incorporate the physical sciences and the life sciences, civilian research and defense.

Bush's distinction between basic and applied research would resonate throughout the next half-century of science funding debate. For Bush, the linear model of fundamental "basic" research in areas without obvious practical application leading to more "applied" research with more obvious military, industrial, or commercial use was a pragmatic construction. To those in Congress and the Administration to whom he would have to sell his plan, the idea that the federal government should be investing in the development of technologies with obvious commercial or industrial use was anathema. However, they could conceive of government support in areas where there was not an obvious commercial application and therefore not likely to be any commercial investment. This was not just supporting science for science's sake. Supporting this "basic" research—research otherwise not undertaken—would start the science down a path that would ultimately lead to deployable innovations, innovations he argued that would "bring higher standards of living, [lead] to the prevention or cure of diseases, [promote] conservation of our limited natural resources, and [assure] means of defense against aggression." (Vannevar Bush, *Science, The Endless Frontier*. Read at: *http://www1.umn.edu/scitech/VBush1945.html* [last viewed November 6, 2003] (original cite: Washington, D.C.: U.S. Government Printing Office, 1945)). However, in reality the distinction between basic and applied research is not quite so clean, nor the actual progress of science quite so linear. The cycle of innovation and technological development is often comprised of multiple episodes of "applied" and "basic" research phases. Yet Bush's model of innovation has clearly polarized the debate in Washington for more than 50 years, where the perception has grown that "basic" research is worthy of government support, "applied" research is often not.

Bush's proposal also helped ossify the military and civilian research enterprises into two distinct entities. His plan to incorporate both civilian and military research efforts in the new agency—an agency under civilian control—was a nonstarter with the military services, who would brook no interference in determining the direction of their research efforts. Instead, the military continued funding the civilian research it had begun to support through what would become the Office of Naval Research in 1947. Therefore, Bush's proposed new agency, which would eventually result in the creation of the National Science Foundation (NSF) in 1950, would focus exclusively on civilian basic research. The Department of Defense would become the federal government's largest supporter of research and development through its own labs and agencies—distinct from NSF.

Bush's proposal also did not sit well with the nation's medical research establishment. Rather than be relegated into one portion of a larger research agency, medical researchers argued they should remain distinct as the Public Health Service's National Institute of Health, a small agency that had been in existence in various forms since 1887. This turned out to be remarkable foresight as the life-science focused NIH prospered over the next 50 years—growing to a budget of nearly $25 billion annually in 2002—while the physical-science-focused NSF languished in comparison at less than a quarter the size.

Bush had grand designs for federally supported basic research and the failure of many elements of his plan has had a marked effect on federal R&D policy ever since. Perhaps most significantly, his inability to convince Congress and the Administration of the need to have one agency responsible for the breadth of federal basic research—physical sciences through life sciences, military through civilian—has resulted in the creation of three distinct constituencies:

- civilian natural and physical sciences, supported most clearly through NSF, but also through agencies such as the National Aeronautics and Space Administration (NASA), the National Oceanic and Atmospheric Administration (NOAA), and the National Institute of Standards and Technology (NIST);

- civilian life sciences, supported by NIH; and

- defense science, supported through the military services and defense-wide labs.

The practical effect of the split is both a competition for limited discretionary funding and a lack of coordination across constituencies, leading to concerns about duplication of effort and other inefficiencies. The latter effect is compounded by the organization and jurisdiction of the congressional committees who oversee the three constituencies. Each is represented by a different set of authorizing and appropriations committees in Congress. The natural and physical sciences, for example, fall largely under the authorizing jurisdiction of the Committee on Science in the House of Representatives and the Committee on Commerce, Science, and Transportation in the Senate. The life sciences find their authorizing jurisdiction under the Committee on Commerce and Energy in the

House and the Committee on Health, Education, Labor and Pensions in the Senate. Defense-related R&D is overseen by the Armed Services Committees of both the House and Senate. On the appropriations side, the natural and physical science agencies generally (with some exceptions) find their appropriators in the VA (Veteran's Administration)-HUD (Housing and Urban Development)-Independent Agencies subcommittee, the life sciences in the Labor-HHS subcommittee, and the defense researchers in the Defense appropriations subcommittee. Coordination between any two of these congressional committees is difficult and somewhat rare. Coordination between all of them is near impossible. Had Bush been successful in convincing Congress and the Administration that a single agency represented the best model for the federal research enterprise would coordination and priority-setting be as contentious as the current arrangement? It is impossible to know, and it is unlikely to ever change. Committee jurisdictions are hoarded and defended fiercely by committee members. It is hard to imagine the situation in which members of Congress would voluntarily cede oversight of an agency. The creation of the Department of Homeland Security in 2002 out of the pieces of a number of different agencies after the extraordinary events of September 11, 2001, required the creation of three new congressional committees and the largest jurisdictional shakeup in 50 years.

The creation of large, cross-agency programs such as NITRD is largely the result of this diffusion of oversight responsibility for science agencies throughout the legislative branch. Through the establishment of an executive branch-based National Coordinating Office—a significant component of the NITRD program—Congress and the Administration have attempted to mitigate some of the difficulties encountered when each agency involved in funding information technology research has its own budget process and its own congressional committees with which to contend. One thing NITRD does not do, however, is create a monolithic funding organization for IT-related research. The agencies of NITRD are a quilt of different funding models, missions, and philosophies—a fact many would argue is a strength of the current arrangement. Research funded by NIH tends to evolve differently, for example, from research supported by the Defense Advanced Research Projects Agency (DARPA). NASA funds IT R&D related to its aeronautics and space mission, while NSF takes a much broader view of the research enterprise. These agencies, and their approach to IT R&D funding, were shaped by the conditions that existed at their founding, which for many was immediately after Vannevar Bush's proposal.

Early Computing Research

Digital computing was truly in an embryonic stage at the end of World War II (WWII) in 1945. In 1943, the Army Ballistics Research Laboratory had contracted with the University of Pennsylvania to produce an Electronic Numerical Integrator and Computer (ENIAC) to calculate ballistic trajectories for ordnance. By 1945, ENIAC's 19,000 vacuum tubes and 1,500 relays were fully assembled and functioning, becoming America's first digital electronic computer. Like most of the computers constructed over the following 15 years, ENIAC was put to use by the federal government—in its case, by the U.S. Army Ordnance Corps. In fact, surveys through 1961 of all electronic stored-program computers in use in the country reveal "the large proportion of machines in use for government purposes, either by federal contractors or in government facilities." The federal government was both the primary funder of computing technologies and the primary end user.

Three agencies were particularly critical at this early, post-WWII stage. The U.S. Navy's Office of Naval Research (ONR) funded much of the theoretical mathematics work that underpinned digital computing. Just as importantly, ONR sponsored numerous conferences and publications as a way of building up the intellectual infrastructure for computing. ONR also funded computer construction for other federal agencies, including machines at the National Bureau of Standards and Massachusetts Institute of Technology (MIT). The National Bureau of Standards (NBS)—later NIST—was both a consumer of computing technology, contracting for the first three commercial, electronic, digital, stored-program computers (Univac), and a builder of its own systems. Finally, the Atomic Energy Commission (AEC) was a prime builder and purchaser of computer systems to aid in the development and testing of nuclear weapons. The Los Alamos National Laboratory built its first computer in the late 1940s (called MANIAC) and Argonne and Oak Ridge National Laboratories both soon followed suit. It is also notable that AEC purchased commercially built computers as well, serving as launch customer (to borrow a term from the aviation industry) for IBM in the early 1950s and Remington Rand in 1955.

This period also saw the creation of a number of important private organizations that sprang from governmental participation. In 1946, the Air Force and Douglas Aircraft joined forces to create what would become the RAND Corporation to study new methods of warfare. RAND soon separated from Douglas and became an independent, non-profit think tank for first the Air Force, then other defense and civilian agencies, largely responsible for training hundreds of early computer programmers. The Lincoln Laboratory at MIT also got its start as a federally funded research and development center in 1951, chartered to study air defense systems and ways to protect the nation from possible nuclear attack. MIT, which already had experience building its own "Whirlwind" computer system in the late 1940s, focused its efforts on building a network of ground-based radars that would employ computer processing of radar data for the first time.

In this embryonic period, government funding was the defining characteristic of nearly every computer effort. Even revolutionary work on the development of the transistor taking place in the privately run Bell Labs (now a part of Lucent) was supported in part by government investment. Yet lack of coordination among

programs led to accusations that the effort "was open to criticisms of waste, duplication of effort, and ineffectiveness caused by rivalries among organizations and
their funding sources." Nevertheless, world events would quickly give scientists
and policy makers new motivation for increased scientific investment and new and
reshaped science agencies.

Sputnik and the Intensifying Cold War

The landscape of U.S. science was fundamentally altered by the October 4, 1957,
launch of the Sputnik satellite by the Soviet Union, the world's first artificial satellite. Orbiting the Earth every 98 minutes, this basketball-sized metal sphere served
warning to an already wary U.S. public that Soviet science had leapt ahead of U.S.
efforts. The launch of Sputnik, followed one month later by a more ambitious
launch that put a dog in orbit on Sputnik II, demonstrated that the Soviets had the
technological wherewithal to put payloads in orbit and, therefore, perhaps had the
ability to launch missiles from Europe to the United States carrying nuclear
weapons.

The political response to this escalation of the Cold War and resulting "crisis of
confidence" in American capability was quick. Shortly after Sputnik I's successful
launch in October 1957, the Department of Defense, which had an existing satellite program through ONR called Vanguard in the works since 1955, redoubled its
efforts by authorizing an Air Force program called Explorer on an accelerated
schedule. Explorer I's successful launch on January 31, 1958, marked America's
official entry into what became "the Space Race" with the Soviet Union, one long
battle of the Cold War fought with science and technology rather than soldiers and
tanks. Determined not to be caught flat-footed again, in February 1958, the
Department of Defense (DOD) established the Advanced Research Projects
Agency (ARPA), with the idea that it would become the department's "space
agency." However, when Congress established the National Aeronautics and
Space Administration (NASA) a short five months later, ARPA's focus shifted
from space to assuring that the United States maintained "a lead in applying state-
of-the-art technology for military capabilities and to [preventing] technological
surprise from her adversaries."

Many aspects of ARPA's original structure made it unusual within DOD. To
avoid interservice rivalries, the agency director reported directly to the Secretary
of Defense, rather than through any particular branch of the military (unlike
ONR or the Air Force Research Labs). Its staff was composed of a small core of
managers supported by a corps of scientists and engineers who rotated through
the agency on short (usually two-year) assignments and who looked to fund
long-range, high-risk research at existing agencies and academic institutions
rather than building labs. The result was that the agency was remarkably nimble

for a federal entity, able to act on and approve new research programs with minimal bureaucracy.

ARPA established itself as crucial in the advancement of computing research beginning in 1961 under the leadership of director Jack Ruina. Ruina, a researcher from the University of Illinois, was the agency's first "scientist-director" and saw great value in decentralizing management at the agency in order to allow the directors of the various research offices to exercise relative independence in managing their research programs. Also importantly, Ruina saw great value in ARPA pursuing the cutting edge of research areas, even if that research was not immediately relevant to military uses.

For computing, Ruina's most important act was the establishment of the Information Processing Techniques Office (IPTO) in 1962 and his appointment of J.C.R. Licklider, a Harvard-trained psychologist, to head it. Licklider had studied computing research extensively and brought to IPTO a clear vision to focus on the "man-computer symbiosis" to bring together the "then-disparate techniques of electronic computing to form a unified science of computers as tools for augmenting human thought and creativity." (National Research Council, Computer Science and Telecommunications Board, *Funding a Revolution: Government Support for Computing Research*. Washington, D.C.: National Academy Press, 1999, 99). Licklider's IPTO moved computing research from thinking about how to program the new, powerful digital machines to how to use this "tremendous new power…for other than purely numerical scientific calculations."(100)

Licklider and the IPTO directors who followed him throughout the 1960s and subscribed to his philosophy were responsible for ARPA advancing the science and technology of computing more than any other federal agency, and soon had a budget to show for it. IPTO's 1965 budget of $15 million dwarfed many other federal research agencies. Research funded by IPTO during this time helped push developments in time-sharing, interactive computing, computer graphics, and software that ultimately moved computing from their strictly business or military use to the ubiquitous uses today.

In addition to having the right people in charge at the right time, a significant reason for ARPA's relative success in pushing computing forward had to do with the way it structured its research funding effort. The agency had (and still has) a number of freedoms in funding research that agencies such as the National Science Foundation did not (and still does not). ARPA program managers such as Licklider could fund particular "centers of excellence"—typically university research departments such as Stanford or MIT—without concern for "equitable" distributions of funding based on geography or any other factor beyond scientific capability. Nor was ARPA research peer-reviewed in the same way that competitive grants at NSF were evaluated. ARPA program managers had great discretion to fund projects they believed to be promising—as block-grants or long-term, multi-year efforts—without a formal, sometimes time-consuming, peer-review process. In this way, ARPA was able to create and nourish communities of researchers to focus on problems of particular interest to the agency, with great success. The key, according to National Research Council study of the period, was the "light touch" exercised by ARPA management, which greatly reduced red tape and allowed program managers and researchers to concentrate on research.

At the same time, NSF was also supporting individual computing researchers at a wide range of institutions, helping legitimize computer science as a discipline worthy of independent study. Just as importantly, NSF was investing heavily in computing infrastructure at American universities, insuring that a wide variety of researchers had access to these new, powerful tools. This period also marks the rise of computer science departments, notably Stanford and Carnegie Mellon in 1965 and MIT in 1968.

The computing industry also grew dramatically during the 1960s. IBM, Burroughs, Control Data, GE, Honeywell, NCR, RCA, and Sperry Rand formed the core of this new industry that had grown large enough that some companies could support their own—in some cases significant—research and development efforts. IBM's Watson Center, opened in 1961, was a notable early example. The establishment of industrial labs helped refine the focus of federal research efforts. The industrial labs were geared toward transitioning basic research to applied or developmental research that could one day lead to commercial technologies. These private labs relied upon a strong, truly basic research effort supported by the federal government at universities and federal labs as the source of much of their research, and therefore helped make the case to federal policy makers that supporting basic research was an appropriate and necessary role for the government.

Though the industrial labs' existence seemed to make the case that private enterprise was best-positioned to support applied research, that fact was not unanimously accepted by those in the scientific community during this period. In fact, the issue became part of a suite of issues that motivated many previously non-partisan scientists and engineers to use their credibility as men of science to political ends during the 1964 presidential campaign, in what represented the first (and so far, last) foray of a significant group of researchers into partisan politics.

Scientists and Engineers for Johnson-Humphrey

In the presidential elections prior to 1964, America's pre-eminent scientists and engineers generally played a small, relatively non-partisan role. Where there was involvement, it was generally as a member of a candidate's "brain trust" on science and technology issues, a position generally seen by both the candidates and scientists themselves as non-partisan. In large part this was because the issues of science did not generally break down along party lines. Support for federal funding for basic research was championed by both Democrats and Republicans alike. The issue of federal support for applied research did (and does) tend to break along party lines—Democrats tended to be in favor of supporting research that would lead to immediate or near-term improvements to American industry, while Republicans generally believed that the federal government should not support research that private industry could reasonably be expected to under-take—but the consequence of the break was not so significant as to lure many scientists out of their non-partisan world.

However, the election of 1964, which pitted Democratic President Lyndon Johnson against Republican Senator Barry Goldwater, changed this calculus. In a world increasingly wary of the real threat of a nuclear exchange between the Soviet Union and the United States, Goldwater had made waves by suggesting the Administration was not taking a hard enough line with the Soviets. Goldwater was seen by many, fairly or unfairly, as a "cowboy," a self-avowed "extremist" whose shoot-from-the-hip comments (he once joked about lobbing a missile into the men's room at the Kremlin) could scare the Soviets and make the world a much more dangerous place. Among those sharing this view were a large number of the nation's most pre-eminent scientists from across the political spectrum, including George B. Kistiakowsky, President Eisenhower's science advisor, and Detlev Bronk, former president of the National Academies of Science.

The two scientists were among those who joined a growing organization called Scientists and Engineers for Johnson-Humphrey, a group aimed not at advocating issues of science policy, but at depicting Goldwater as "unsuitable for the nuclear-age presidency, unthinkable for managing the dangerous standoff between the United States and the Soviet Union." (Daniel S. Greenberg, *Science, Money, and Politics: Political Triumph and Ethical Erosion*. Chicago: The University of Chicago Press, 2001, 153) Begun with $12,000 in seed money from the Johnson campaign, the organization soon grew to a self-sustaining organization with a membership of nearly 50,000 scientists and engineers. The organization ran over 100 newspaper ads and over 3,000 radio ads in support of Johnson-Humphrey and against Goldwater in the weeks leading up to the election. Perhaps most damag-ingly, they sponsored a radio roundtable hosted by the former director of DOD Defense Research and Engineering under Eisenhower and featuring "men from science, engineering and medicine,"—including Harold Urey, Nobel laureate and member of the Manhattan Project; Admiral W.F. Raborn, who worked on the Polaris submarine missile system; and Kristiakowsky—who had all "helped develop the power that could destroy mankind" (Greenberg 155) and hoped to make sure Goldwater was not put in a position to use it.

The broadcast was devastating for Goldwater. Denison Kitchel, who ran Gold-water's campaign, said later: "My candidate had been branded a bomb-dropper, and I couldn't figure out how to lick it." (Greenberg 157) Johnson was reelected, garnering 61 percent of the popular vote.

Many of the scientists involved in the campaign had great misgivings about the role of scientists in future elections however. Many felt that using their scientific credibility for political aims, ultimately undermined that credibility in the long term. In 1968, Philip Handler, who had been active in Scientists and Engineers for Johnson-Humphrey and who was now president of the National Academies of Science and chair of the National Science Board, was asked by Humphrey's campaign staff to organize a similar effort for the 1968 presidential election. After initially accepting, Handler changed his mind and refused, noting that if scientists organized and joined in partisan politics "it is inevitable that national attitudes and federal support for science must also come to involve political considerations." (Greenberg 161)

Twenty Years of Explosive Growth

Despite a still-simmering Cold War, by the early 1970s, public attitudes about federally funded science were changing. An increasingly anti-establishment attitude, stirred primarily by escalating U.S. involvement in Vietnam and increasing concerns about man's impact on the environment, tempered outward support for government-sponsored research. As the costs of the war in Vietnam put a significant squeeze on federal science budgets, the voices of the champions of science did not carry quite so far, and programs felt the pinch. DOD funding for mathematics and computer science reached a two-decade low in 1975, and President Nixon pushed NSF out of the business of developing computer networks, in favor of private industry solutions.

There was a new, shorter-term funding perspective as well—a focus on insuring the taxpayer dollars were spent in ways most obviously beneficial to the nation. An amendment to the 1970 Defense Authorization bill called the "Mansfield Amendment" forbade military funding for research that did not have a "direct or apparent relationship to a specific military function" (NRC 112) was emblematic of this thrust. Though the amendment was not readopted in subsequent years, it colored the perception of federal funding for some time and had the long-term consequence of focusing agencies like ARPA—renamed the Defense Advanced Research Projects Agency (DARPA) in 1972—increasingly on shorter-term research efforts, or efforts with more obvious military application. The act also signaled to DARPA that a change in the type of research it supported was required. Traditionally, DARPA had supported what the DOD classified for accounting purposes as 6.1 research—truly basic research, as opposed to 6.2 (Applied Research) or 6.3 (Advanced Technology Demonstration)—in addition to a significant amount of 6.2 (Applied) research. After the Mansfield Amendment, DARPA clarified its mission somewhat to reflect its role as a "bridge" between 6.1, 6.2, and 6.3 research. As a result, DARPA's role as a supporter of 6.1 research diminished. This was especially true in IPTO, which at its outset under Licklider had funded exclusively 6.1 research. By the early 1970s, the emphasis had shifted significantly to 6.2, which at that time constituted over half the IPTO budget. The debate over DARPA's role funding 6.1 and 6.2 research continued in the years leading up to the president's FY 2003 budget, but the situation remained relatively unchanged. For FY 2003, DARPA estimated that less than 5 percent of its budget would be spent on 6.1 research.

The shift to more applied research at DARPA did have its positive effects, however. ARPANET, the precursor to today's Internet, was funded during this period to demonstrate networking research worked out in the 1960s. Robert Kahn, who joined DARPA as a program manager in 1972, embraced the applied model and used it to push forward research in packet radio, networking, and Internet-working protocols, including TCP/IP, the protocol that became the basis of the Internet.

Kahn became head of IPTO in 1979 where he advanced two key programs with long-ranging effects. First, building on pioneering work on very large scale integrated circuits (VLSI) in the mid-1970s by Carver Mead at CalTech and Lynn Conway at Xerox PARC, DARPA supported work on a number of innovations that revolutionized computing and computing research. The DARPA VLSI program attempted to identify and support ongoing research in computer workstations, reduced instruction set computing, and semiconductor fabrications services for university researchers in an attempt to accelerate development of a broad set of improvements to computer capabilities. Some of the fruits of this research—and there were many—were quickly adopted by the industry, or were spun off by the researchers themselves. Development of a single-user workstation under a DARPA grant became the basis for the Stanford University Network (SUN), which combined with a version of Unix funded under another DARPA VLSI contract, and graphics work ongoing at Stanford, became Sun Microsystems. For its part, Sun Microsystems licensed the RISC technology developed at UC Berkeley under VLSI. Just as importantly, VLSI supported the development of the Metal Oxide Silicon Implementation Service (MOSIS), a method of abstracting computer chip design so that university researchers could quickly manufacture limited numbers of custom or semicustom chips at reasonable cost.

The program was also important as a demonstration of DARPA's ability to foster revolutionary advances by nurturing research communities. DARPA's flexible funding regime allowed for funding in a broad range of areas, some with only indirect connections to military or defense applications. Also important for information sharing was the maintaining of open, nonrestrictive policies on publication of results. For DARPA, and the researchers working on these problems, it was important that the work not be classified so that the community might benefit from sharing it. There was, however, a general understanding that in order to benefit U.S. industry over foreign competitors, research results would be shared immediately only within the community of VLSI researchers and not published for at least a year.

Kahn's second program was called the Strategic Computing Initiative (SCI), assembled in part to combat America's perceived loss of leadership in the semiconductor and computing industry to the Japanese. In the early 1980s, after having already suffered the loss of American dominance in the automobile industry to the Japanese, Congress and the Administration were receptive to the argument that a new thrust in computing research aimed at microelectronics, supercomputers, generic applications, and defense applications was necessary to promote a strong domestic electronics and computer industry (which, in turn, was critical to national security). The SCI program, which had the overarching goal of improving artificial intelligence systems, was not primarily based in university-led research. Rather, nearly half of SCI's research funding went directly to industry, "with corresponding emphasis on tangible results and applications," according to a National Research Council study. The program was indicative of the shift at DARPA away from fundamental research toward research capable of producing demonstrable results over short time frames.

The debate over the value of this shift in focus from relative long-term to short-term research at DARPA would reemerge as part of the FY 2003 appropriations process. Though university-based computing researchers had grown concerned for some time about what they perceived as a continuation of this shift toward funding more short-term, focused research, the issue did not come to a head until DARPA director Anthony Tether took over the agency in 2001. Tether, an electrical engineer with a long career in industry, was plain in his desire to reshape DARPA in the model of a high-tech venture capital firm—identifying promising technologies early and providing them with the capital needed to turn them into demonstrable technologies on short timelines. Key to this identification process was Tether's implementation of a formal "go/no-go" decision matrix for all DARPA-funded research projects. In addition to facing a traditional annual review, in which DARPA managers would verify that contract work was proceeding according to plan and budget, DARPA contract recipients would now face multiple review milestones at relatively short 12- to 18-month intervals, by which their projects must deliver some demonstrable result in order to receive continued funding.

To the Administration, DARPA's approach appeared to represent a reasonably business-like approach to providing good stewardship over taxpayer dollars in the course of developing the technologies necessary for national security in the post-September 11th world. However, for university researchers accustomed to working on basic research problems, the idea of "scheduling" breakthroughs or demonstrable results on 12-month timelines was anathema to the basic research enterprise and nearly impossible to do in an academic environment. Finding little support for change within the Administration, the university researchers took their concerns to Congress during the FY 2003 appropriations debate, where they made the case that DARPA's new funding regime had essentially prohibited university researchers from pursuing DARPA contracts, effectively preventing some of the best minds in the country from working on national security problems. The "go/no-go" decisions, they argued, would result in research that was evolutionary, not revolutionary, with potential grantees only proposing ideas for which they were sure to deliver significant progress in 12 months. Failing to consider long-term research could leave the nation once again "flat-footed" to the new threats of the 21st Century.

The researchers got a sympathetic reception from members of Senate Armed Services Committee who put the question to Tether during his annual appearance before the committee as part of the Defense Authorization process. Tether brushed the concerns aside in his official response:

> I do not believe our "Go/No-Go" milestones will make our work less revolutionary nor do I think they will interfere with university participation in our programs. Instead, I view them as a technique for providing solid management and accountability for the significant investments we make with taxpayer dollars. … This technique allows progress to occur quickly and keeps everyone focused on accomplishing goals they can see happen yet that will still have a big long-term impact. Industry understands this method because it is a technique used by the best industrial managers for executing a difficult multi-year contract.

The Committee remained unconvinced, however, and included an admonition to DARPA in the committee report accompanying FY 2003 Defense Authorization Act:

> The committee, however, is concerned about recent trends in the agency-sponsored research that appear more shortsighted in their approach, particularly the emphasis on 12- and 18-month reviews in order to attempt to eliminate non-promising technologies.
>
> The committee supports effective internal oversight and commends DARPA for pursuing truly innovative technologies. However, annual reviews may not be appropriate for all basic and applied defense-related research programs. Additionally, these reviews have a discouraging effect on the intended long-term payoff of the research and are especially inconsistent with the time frames and pace of university research. The committee is concerned that this near-term approach to basic and applied research will have detrimental consequences on the ability to develop innovative solutions to future threats. Therefore, the committee urges DARPA to re-evaluate its policies for reviewing and terminating awards in scientific and technical areas where the Department of Defense is dependent on

DARPA's ability to do revolutionary research that requires some time to develop and mature.

At the same time DARPA's efforts were initially turning more toward applied-oriented research, NSF's computing science programs were slowly increasing in prominence and budget. Between 1973 and 1985, NSF's budget for computing research quadrupled. By 1977, NSF was the largest funder of basic research in computer science. In 1986, NSF created the Computing Information Science and Engineering Directorate (CISE) and named Gordon Bell, formerly of Digital Equipment Corporation, as its director. The new directorate was important both symbolically and fiscally. Symbolically, the creation of CISE was the result of the recognition that computer science had achieved the same status as the other sciences in NSF's pantheon of directorates, including biological sciences, mathematics, and physical sciences. Fiscally, incorporating all computing research under one directorate allowed it to achieve a relative budget importance that was not possible in the scattering of locations it had occupied previously. By 1987, CISE had grown to $117 million and represented over 7 percent of the Foundation's total budget.

Concerns about American economic competitiveness also helped to dramatically reshape the funding landscape in the late 1980s. Building on the arguments that had made DARPA's SCI program possible, and under threat from growing international competition, 14 U.S. semiconductor companies joined in a non-profit consortium to pursue new strategies for production of next-generation semiconductor devices. The consortium, called SEMATECH, gathered the support of Congress as well, which authorized DARPA to match the semiconductor industry contributions of $100 million a year for five years in order to promote the partnership. Under the partnership, which ran until 1995, SEMATECH set industry specifications, funded R&D at supplier companies to improve production techniques, and improved links between the semiconductor manufacturers and their suppliers. The legislation that led to the SEMATECH/DARPA partnership also formalized these types of government/industry relationship into new entities called Cooperative Research and Development Agreements (CRADAs). Since then, nearly every federal research agency has established CRADAs with industry partners in support of advancing agency missions and increasing the flow of innovation into American industry.

SEMATECH stopped receiving federal funds in 1995 in order to focus on its own research unhindered by federal requirements. The organization remains a viable, industry-supported research consortium that currently includes all the major semiconductor manufacturers in the country, including Intel, AMD, IBM, Motorola, and Texas Instruments.

Bayh-Dole Act

As federal funding for research ramped up during the 1970s in computer science and elsewhere in the R&D portfolio, long-standing concerns among policy makers that U.S. industry was not seeing the real benefit of the taxpayer's investment in research received more attention. Federal funding was providing copious amounts of quality research, it was argued, but U.S. companies were not often enough able to capitalize on it. Companies wishing to commercialize a product developed with federal funds at a university, for example, faced a web of tight restrictions on licensing, varying patent policies among agencies, and lack of exclusive manufacturing rights for government-owned patents. In 1980, according to the Association of American Universities (AAU), only 5 percent of government-owned patents resulted in new or improved products. In response, Congress passed the Bayh-Dole Act, which allowed universities and other non-profit organizations to own patents for federally funded inventions they produced.

Proponents of the act believe that the act would promote technology transfer by providing incentives for researchers to consider the commercial potential of their inventions and for the universities to seek industrial partners who might benefit from them. This has been borne out in practice. The AAU notes that prior to Bayh-Dole, universities were granted fewer than 250 patents a year. In 1996, universities "received more than 2,000 new patents, executed nearly 2,200 licensing agreements, and received royalty income from licensing of $242 million. Since 1980, more than 1,500 start-up companies have been formed based on technologies discovered at academic institutions." (Association of American Universities, "University Technology Transfer of Government Funded Research," web site, last viewed 12/15/03, at *http://www.aau.edu/research/TechTrans6.3.98.html*)

Critics of the act, such as former Commerce Secretary Phillip Klutznick, who testified against the bill in Congress, note that allowing private companies to capitalize on promising research paid for with taxpayer funds is like "using tax money to pay a contractor to build a road and then allowing the contractor to charge an additional toll to those who travel the road." (Quoted by Judith Gorman, "Paper Cuts: The Golden Fleece", June 13, 2000, Web article. Published by AlterNet. Last viewed 12/15/03 at *http://www.alternet.org/story.html?StoryID=9290*) The Act, critics say, creates an implied duty to commercialize, which can also be harmful to notions of academic "purity," and foster fears that researchers will shade their results if they have a financial stake in them.

The Act also has an indirect effect on university computing science researchers in a way that the bill's authors might not have anticipated. Recognizing the real benefits to both the computing community and, indeed, the world as a whole that have come from the free release of software code for operating systems such as BSD Unix and the TCP/IP Internetworking protocol, some computer scientists have raised concerns that university technology managers, focusing on the commercial potential of software developed under research grants at their universities, will bring pressure to bear on researchers to not share their code, lest they give away valuable intellectual property rights. "I don't know whether they would let us release software like TCP/IP today," Susan Graham, a professor of computer science at Berkeley, told Salon.com. (Jeffrey Benner, "Public money, private code." Salon.com Website. January 4, 2002. Page 3. Last viewed 12/15/03 at *http://archive.salon.com/tech/feature/2002/01/04/university_open_source/index.html*) "If they thought it had monetary value, they would want a revenue stream. There would be companies who could pay for it. I'm not sure we would have the same outcome [as in the past], and that's what concerns me." (Quoted in Greenberg, 210)

The '90s and the New Century: NITRD Takes Shape

Computing research by the late 1980s had grown substantially from its low point in 1975. Federal spending on basic computer and computational science increased from less than $40 million per year in 1975 to over $230 million [inflate/deflate] spread over 10 different federal agencies in FY 1990, and total computing research and development (basic and applied) had grown to nearly $600 million. With so many agencies spending so many dollars on research, concerns about waste, duplication of effort and other inefficiencies led Congress to formalize a method of inter-agency coordination. In 1991, Congress passed the High Performance Computing Act, creating the High Performance Computing and Communications Initiative (HPCCI), a 10-agency working group that would coordinate basic information technology research and development across the federal government.

At the same time, the single most enduring external influence on federal funding—the Cold War—was coming to an end. The opening of the Berlin Wall in 1989 was soon followed by the collapse of the Communist dictatorships of Eastern Europe, the reunification of Germany, and ultimately, the disintegration of the Soviet Union in 1991. With the end of the Cold War, federal spending priorities began to shift in ways that significantly reshaped the federal R&D landscape. Without the spectre of a powerful adversary in the Soviet Union, politicians in both parties moved quickly to reduce overall defense spending, which in turn reduced overall defense-supported R&D. Between FY 1990 and FY 1995, Department of Defense-sponsored research funding dropped 22 percent.

To many in the science establishment, the end of the Cold War meant losing the most obvious justification for science funding. Former House Science Committee Chairman George E. Brown Jr. lamented as late as 1998: "The end of the Cold War disrupted the link between science and a clear justification for its support, and has meant shrinking budgets in some areas of research." (National Science Board Commission on the Future of the National Science Foundation. "A Foundation for the 21st Century: A Progressive Framework for the National Science Foundation." NSB92196. Last Viewed on 12/15/03 at *http://www.nsf.gov/pubs/stis1992/nsb92196/nsb92196.txt*) The U.S. in 1991 was also in the throes of an economic recession, and questions were being raised about the role of federally funded research and development—both basic and applied—in helping American industry, and thus the economy, recover. Japan and Europe seemed to be becoming economically and scientifically dominant, making "competitiveness" a key concern in Washington. To policy makers, Japanese industry in particular seemed to be much more effective than U.S. firms at turning technological innovations—usually innovations originating outside of Japan—into marketable products, and the trends in high-tech imports and exports seemed to support that belief.

As a result, the appropriate balance between federally supported basic and applied research once again became an issue confronted at most federal science agencies, particularly at the National Science Foundation, the nation's de facto source of basic research funding. Though there remained an ideological divide between Republicans and Democrats concerning the propriety of the federal government investing in research obviously beneficial to U.S. industry—Republicans generally believed that the federal government only had a legitimate role funding truly basic research; whereas, Democrats generally believed there was an obvious and relatively immediate benefit to industry from strategic federal investments in particular technologies—increasing pressure was put on NSF by both sides in the early 1990s to consider broadening its role as a funder of research. NSF was asked to consider whether forming closer ties with industry was an appropriate expansion of its mission, and a valuable service to the nation's economy.

At the same time, Democrats in Congress also pushed through increased funding for a relatively new federal program (begun in 1988) run by NIST called the Advanced Technology Program. The aim of ATP was direct funding of promising technology research at U.S. companies. Republicans labeled it corporate welfare and lamented the federal government's involvement in "industrial policy," but facing the economic downturn of the early 1990s, President George Bush acquiesced to Democratic demands and allowed funding for the program.

These developments did not go unnoticed by the scientific establishment, especially those involved in university-based research. Researchers in academia railed against any proposed change to NSF's mission for fear that it would dilute support for basic research at universities. They found an ally in a special commission established by newly appointed NSF Chairman Walter Massey to evaluate the future of the NSF. After 75 days of consideration, the commission concluded that NSF's emphasis on basic versus applied research was not to blame for America's apparently lackluster standing in the world economy. Instead, America's failures in the marketplace were due to the competition's "shorter product cycles, lower costs, and superior quality." The report concluded: "All manner of other more prominent factors, including the stewardship by American business, far outweigh whatever could be traced to the technology itself or the technologists." (Semiconductor Industry Association, "About the SIA." Association website. Last viewed 12/15/03 at *http://www.semichips.org/about.cfm*) As a result, NSF invoked the original plan of Vannevar Bush and did not broaden its mission to embrace a role as a supporter of applied research—a position it maintains today.

However, the end of the Cold War also meant other changes for the federal science portfolio. The early 1990s also saw an increasing emphasis on health science research, predominantly through the National Institutes of Health. NIH had always received bipartisan support for its support of medical research, but a shrinking defense budget and high-profile efforts aimed at curing cancer and AIDS research provided members of Congress with the wherewithal to begin a significant effort to increase funding for the agency. From FY 1990 to FY 1995, Congressional appropriators increased NIH funding by more than half. By FY 2003, NIH's budget had more than doubled to nearly $25 billion. At the same time, NSF's budget was growing at a much more modest pace, and by FY 2003 had hit the $5 billion mark. While this was generally good news for those in the life sciences, researchers in the physical sciences began to chafe at the growing imbalance.

The physical scientists argued that advancement in the life sciences was heavily dependent upon fundamental work and innovation in the physical sciences and therefore shortchanging the physical sciences would have serious long-term effects on the pace of innovation in the life sciences. Developments such as Magnetic Resonance Imaging (MRI), the use of lasers, and sophisticated computer models and visualizations that had proven so valuable in health research, they argued, were made possible by fundamental research in the physical sciences. In 2000, NIH Director Harold Varmus agreed, noting that much of the most exciting and promising new research performed at NIH was heavily dependent on disciplines traditionally provided by the NSF: computer science, chemistry, physics, and engineering.

This issue also came to a head during the FY 2003 budget deliberations. In 2002, the President's Council of Advisors for Science and Technology (PCAST)—a panel of pre-eminent members of the science and technology community, both in academia and industry—was tasked with evaluating the state of the federal research portfolio. In August of 2002, they reported that, though increasing investments in the life sciences were still justified, there was a clear need for further investment in the physical sciences.

Though the PCAST report was careful to avoid recommending a specific funding level and did not single out NSF as specifically in need of significant increases, supporters of NSF in Congress focused on these findings as further evidence that NSF deserved the same effort to double its budget as NIH had received beginning five years earlier. By December 2002, Congress had passed and the president-signed legislation authorizing the doubling of NSF's budget by FY 2007. While authorization was an important first step, congressional appropriators still had the final say as to whether Congress would hit those funding targets each year.

Throughout the 1990s, computing research funding continued to grow at a rapid rate. Between FY 1990 and FY 2000, federal support for basic and applied computer science research grew at an average rate of nearly 12 percent per year. Support for the newly created High Performance Computing and Communications Initiative (HPCCI) grew from an initial level of $489 million in FY 1991 to just over $1.5 billion in FY 2000.

The coordinated structure of the HPCCI was certainly one factor influencing the continued increases. Having a national coordinating office for the interagency working group allowed for the production of an annual guide to federal IT R&D programs (commonly referred to as the "blue book"). The initiative also merited its own line in the president's budget—a sum total of IT R&D related spending crosscut from all the participating agency budgets. The single line, instead of multiple references to IT R&D buried in each agency's own budget justification, no doubt raised the prominence of IT R&D in the president's budget process and in Congress.

However, more significantly, it was becoming clear to policy makers and citizens alike that computing and networking technology were in the process of revolutionizing industry, health care, the military, and communications worldwide. By the early 1990s, NSF, which had been given the responsibility of managing the rapidly growing Internet, began the process of privatizing its operation. The development of the Web, made possible by the development of Hypertext Markup Language (HTML) in 1991 and Hypertext Transfer Protocol (HTTP) in 1992, and then ultimately the Mosaic web browser in 1993, brought hundreds of thousands, then millions of computer users onto the Internet in short order. By 1995 all commercial restrictions on the Internet had been lifted and new businesses began to grow in the virtual space.

The growth of the Internet seemed to have at least two obvious impacts. First, it was presumed a given by policy makers, despite not yet actually having economic data to prove it, that the Internet and new computing and networking resources would revolutionize American business by streamlining operations and increasing overall productivity. Second, the relative ease of use of the web and the wide variety of information and entertainment it soon contained brought large numbers of Americans in contact with computers and networks, many for the first time. Computing researchers who may have had a difficult time explaining the value and impact of their work when dealing with policy makers in the past now were talking with people who most likely had already experienced the web, or had their own email address (or even their own web page). In the same way it was intuitive to members of Congress that support for NIH and the life sciences meant better health care for Americans, it became equally clear that the economic boom of the late 1990s was largely the result of advances and innovations in science and technology in general, and specifically in computing and networking technologies. From that starting point, computing researchers had a shorter distance to navigate to make the case that basic computing research was crucial to that innovation.

Basic Computing Research Necessary for President's Tax Cut

An excellent example of how the computing research community was able to use the belief that IT was largely responsible for the economic boom times of the late 1990s in order to argue for more support for basic computing research surfaced during the debate over the FY 2003 budget numbers. In an effort to jumpstart an economy clearly no longer in boom times, President Bush proposed an across-the-board tax cut with an estimated cost to the federal government of about $1.6 trillion over 10 years. Over the same period, the Congressional Budget Office (CBO) estimated the federal government would run a cumulative surplus of $5.6 trillion, demonstrating there was clearly enough in the surplus to "pay" for the president's cut. That the track record for accuracy in these types of predictions over the period of the projection is lackluster at best is beside the point. In Washington these were the two numbers that formed the basis of the debate at that moment in time.

In studying those numbers, the Semiconductor Industry Association (SIA)—the "premier trade association representing the U.S. microchip industry" (Semiconductor Industry Association, "About the SIA". Association website. Last viewed 12/15/03 at *http://www.semichips.org/about.cfm*)—noticed an important fact about the CBO projection. CBO assumed a 3.1 percent increase in real gross domestic product (GDP) that in turn was "was based on the assumption that information technology trends experienced in the last half of the 1990's continue over the next decade." Semiconductor Industry Association, "(SIA Position on Federal Science: Increase Support of University Research." January 8, 2002. Rev. 2.2. Whitepaper. Page 3. Available at: *https://www.sia-online.org/downloads/fed_sci_position.pdf*) Looking deeper at the assumptions, SIA noted that CBO actually assumed that 0.2 percent of that GDP growth annually would be attributable to productivity gains caused by computer price declines, and attributed those declines to the continuing acceleration in semiconductors. Moore's Law (after Gordon Moore, then at Fairchild Semiconductor) refers to Moore's 1965 observation that the number of transistors capable of being placed on a chip or integrated circuit quadruples every three years due to innovations and the march of technology. Moore's Law is not really a law in the sense that it guarantees enough innovation will occur to keep on pace, but nevertheless, the development of the semiconductor has accelerated at least as fast as Moore predicted. In fact, since 1995, SIA reports that Moore's Law has accelerated to a quadrupling every two years instead of three.

SIA argued that in order for CBO's projections to be accurate, and therefore the tax cut "justified," semiconductor production must continue on pace with

Moore's Law. Failure to do so would cost the United States 0.2 in GDP, or about $232 billion a year, an amount in excess of the cost of the president's tax cut. And the only way to insure semiconductor production continued on pace, SIA concluded, was to insure there was a constant source of innovation in semiconductor design, something heavily dependent on federally supported, fundamental research.

SIA was not alone in making the economic case for supporting basic computing research. The Computing Research Association (CRA) cited remarks by Federal Reserve Chairman Alan Greenspan, who noted that the growing use of information technology has been the distinguishing feature of this "pivotal period in American economic history." (Alan Greenspan, Chairman Federal Reserve Board. "Remarks by Chairman Alan Greenspan: Economic Challenges in the new century. Before the Annual Conference of the National Community Reinvestment Coalition." Washington, D.C., March 22, 2000. Available at: *http://www.federalreserve.gov/boarddocs/speeches/2000/20000322.htm*) CRA also cited the President's own budget request for FY 2003 which noted "about two-thirds of the 80 percent gain in economic productivity since 1995 can be attributed to information technology." (Quoted in Peter Harsha, "Computing Research in the FY 2004 Budget." AAAS Report XXVIII: Research and Development FY 2004, Intersociety Working Group, 2003.)

The Role of PITAC

One more significant factor in the rapid growth of the federal IT R&D effort is the input of the President's Information Technology Advisory Committee (PITAC). As part of the original 1991 HPCC legislation, Congress authorized the president to appoint a panel of non-federal academic and industry members to advise him on the federal government's information technology research and development efforts. In 1997, President Clinton charged his PITAC committee—an august group that included Bill Joy from Sun Microsystems; Vint Cerf and Bob Kahn, two of the "fathers" of the Internet; Leslie Vadasz of Intel; and a number of other luminaries in IT—with evaluating the full breadth of the federal government's IT R&D portfolio. The resulting report, Investing in Our Future, emphasized the "spectacular" return on the federal investment in long-term IT research and development.

However, PITAC also determined that federal support for IT R&D was inadequate and too focused on near-term problems; long-term fundamental IT research was not sufficiently supported relative to the importance of IT to the United States' economic, health, scientific, and other aspirations; critical problems in computing were going unsolved; and the rate of introduction of new ideas was dangerously low. The PITAC report included a series of recommendations, including a set of research priorities and an affirmation of the committee's unanimous opinion that the federal government has an "essential" role in supporting long-term, high-risk IT R&D. This opinion was buttressed by the inclusion of a recommendation for specific increases in funding levels for federal IT R&D programs beginning in FY 2000 and continuing through FY 2004—an increase of $1.3 billion in additional funding over those five years.

The PITAC report generated some movement in Congress toward authorizing significant increases for IT R&D and NITRD agencies. Bills were authored in both the 106th and 107th Congresses that would have authorized funding for NITRD programs at the PITAC-recommended levels. None of those bills managed to reach full consideration by both chambers of Congress. However, PITAC recommended funding levels for IT R&D programs at NSF did find their way into the NSF authorization bill (the "doubling" bill) that was signed into law in December 2002.

Despite the general tenor of support IT R&D received from the Administration and Congress in the wake of the PITAC report, agency appropriations between FY 2000 and FY 2002, which increased $518 million, still fell short of the PITAC-recommended funding. Though the President's budget request of $1.9 billion for NITRD activities in FY 2003 represented a 2.5 percent increase over FY 2002, it still fell $614 million beneath the PITAC funding recommendation for FY 2003. Computing researchers used this fact in making their case to congressional appropriators throughout the FY 2003 budget process.

The FY 2003 Appropriations Process: So How Does It End?

The federal appropriations process is geared to finish prior to the start of the new fiscal year on October 1. It almost never does, and FY 2003 was no exception. The process of turning a bill into law is an enormously involved one—a bill has to pass in committee, then in the chamber in which it originates, then in the committee with jurisdiction in the other chamber, then in the whole of the other chamber, then, if there are differences between the bill as it passed in each chamber, it goes to a conference committee to negotiate over the difference, then the compromise version goes back to each chamber and only then will it head to the president for a signature (or back to the Congress should he decide to veto it). Each year the appropriations committees are expected to send 13 appropriations bills through this process before October 1.

In 2002, the process was hampered further by lengthy consideration of a bill to create a new Department of Homeland Security, which would result in the largest federal reorganization in 50 years. Work and debate on the Homeland Security bill occupied most of the summer of 2002, typically the time used by the appropriations committees to move their legislation to the floors of their respective chambers. By

October 1, 2003, no appropriations bills had made it through the process. Instead, Congress passed a series of "Continuing Resolutions," temporary laws that provided funding for federal activities at the same level of the previous fiscal year, so that the federal government could continue to operate while the debate on the appropriations bills continued. Eventually, the 107th Congress ran out of time. The debate on the appropriations bills was put on hold until after the 2002 elections in November, with both parties hoping a strong showing would determine the direction the bills should take. With Republicans making big gains in the House and the Senate—indeed, taking control of the Senate—the appropriations bills were shelved until the new Congress convened in January. Finally, in February 2003, five months after the start of the FY 2003 fiscal year, Congress passed all the remaining appropriations bills in one fell swoop. More than a year after President Bush presented his flag-bedecked budget for FY 2003 to the Congress, the 544-page Omnibus Appropriations Act for FY 2003 was finally signed into law on February 20, 2003.

Computing researchers fared well under the act. NITRD programs were funded at $1.98 billion for FY 2003, an increase of just over 7 percent above FY 2002. Computing researchers also benefited from two bills passed just before the 107th Congress adjourned: the Cyber Security Research and Development Act, which authorized nearly $900 million in long-term cyber security R&D at NSF and NIST through FY 2007; and the NSF Doubling Act, which authorized funding levels at NSF that would double its FY 2002 budget by 2007. Though both bills were authorizations and not appropriations, they sent an important message that Congress and the Administration recognized the continuing importance of basic research in computing, and throughout the sciences.

For now, it appears computing research will continue to enjoy strong federal support—the key force in shaping the IT sector over the last 50 years—for the foreseeable future. Recent conflicts in Afghanistan and Iraq, as well as smaller actions elsewhere around the world and other actions of which we may never learn, continue to demonstrate to policy makers that science and technology, and IT R&D in particular, are crucial to national security. New evidence from economists, including Harvard economist Dale Jorgenson, are finding empirical evidence for what has been assumed since the mid-1990s: that IT R&D is the driver for the new economy. And researchers from across the disciplines will continue to make the case—through the innovations and breakthroughs that will improve our health, our safety, and our quality of life—of the importance of IT R&D in enabling the sciences.

Further Reading

Vannevar Bush, *Science, The Endless Frontier.* Read at: *http://www1.umn.edu/
scitech/VBush1945.html* [last viewed November 6, 2003] (original cite:
Washington, D.C.: U.S. Government Printing Office, 1945)

Daniel S. Greenberg, *Science, Money, and Politics: Political Triumph and Ethical
Erosion.* Chicago: University of Chicago Press, 2001.

National Research Council, Computer Science and Telecommunications Board,
Funding a Revolution: Government Support for Computing Research.
Washington, D.C.: National Academy Press, 1999.

President's Information Technology Advisory Committee. *Information
Technology Research: Investing in Our Future.* Read at: *http://www.nitrd.gov/
pitac/report/index.html* [last viewed November 6, 2003] (original cite:
Washington, D.C.: National Coordinating Office, February 1999.)

Chapter 2

Telecommunications and Computers: A Tale of Convergence

Steve Mosier

"I think there is a world market for maybe five computers."

Thomas Watson, Sr., 1943

"The Network is the computer."

Scott McNealy, 1982

At the start of the Computer Age, Thomas Watson, Chairman of IBM, said, "I think there is a world market for maybe five computers." Focusing only on the computer, he did not foresee the possibility of combining the processing power of computers with the ability to collect and share data that a modern telecommunications system could provide. Years later, Scott McNealy of Sun Microsystems would say, "The Network is the computer," recognizing that the value of a single isolated computer was dwarfed by a synergistic network of computers. From the time of Watson's statement in 1943 to McNealy's in 1982, great advances in technology were made. Transistors and integrated circuits replaced vacuum tubes, enabling computers to be smaller, less expensive, and faster. During the same period, research in communications theory (particularly at Bell Labs where Claude Shannon published "A Mathematical Theory of Communications" in 1948) began to lay the foundation that would enable the communications industry to take advantage of digital processing circuits. From that point on, the computing and communications industries would be forever intertwined, leading to talk, by the end of the twentieth century, of inevitable "convergence" of the computing, communications, and even entertainment industries. Now, early in the twenty-first century, early adopters are able to download books, music, or movies to their personal computer (the intellectual property issues of this are the subject of another chapter) over a wireless network as they sip mocha cappuccinos at their local café.

Despite the growing technical similarities between computing and communications, the two industries have different origins and developed under different market and regulatory conditions. While one has matured as a regulated public utility, the other has for the most part enjoyed the freedom of the open market system. This chapter explores the growth and regulatory history of the telecommunications industry, particularly as it relates to computing and the combination that has become known as the Information Technology industry, and explores issues and possibilities related to the convergence of these industries.

Telecommunications: Technology, Monopoly, and Regulation

A brief history of technical advances, corporate expansion, regulatory actions, and judicial decisions will help explain how we got to the current state of telecommunications regulations. Telecommunications, defined in *The Telecommunications Act of 1996* as "the transmission, between or among points specified by the user, of information of the user's choosing, without change in the form or content of the information as sent and received," began with the invention of the telegraph. Samuel F. B. Morse received a patent for his version in 1837; government interest in this technology soon followed, and the Telegraph Act of 1843 established a federally operated telegraph line between Washington, D.C. and Baltimore, Maryland. Morse offered to sell the patent rights to the government, but the government declined, so the door was open for private ownership of communications services. Many small regional telegraph companies quickly sprang up, and almost as quickly the merger and acquisition process began, allowing local companies to extend their reach. Western Union soon emerged as the monopolistic winner; its status as the nation's largest corporation by the end of the Civil War prophesied the role that communications would play in the economy from the mid-nineteenth century on.

Western Union maintained its leadership position in long-distance communications with no significant competition until 1876, when Alexander Graham Bell received the first patent for the telephone. Unable to achieve quick commercial success, Bell's patents were offered for sale to Western Union, but Western Union declined to purchase them. Quickly recognizing their mistake and the potential of the voice communications market, Western Union soon bought the patent rights to a rival technology invented by Elisha Gray. Legal battles between Western Union and Bell Telephone soon began, ending with a negotiated settlement in 1879 that would map out the playing field and identify the players of the telecommunications industry until the latter half of the twentieth century. Under the terms of this agreement, each company would drop patent infringement claims against the other; Bell would not enter the telegraph business and Western Union would not enter the telephone business during the terms of the disputed patents. By the time those patents expired in 1893, each company had more firmly established their monopoly positions, and infrastructure and technology costs were effective barriers to other competition. Those companies that did emerge were often in areas not served by Bell. The various Bell companies came together under American Telephone and Telegraph, a Bell company that operated long distance lines rather than local telephone service, in 1899. AT&T was chosen as the holding company primarily because it was headquartered in New York, which had more lenient corporate laws than did Massachusetts, where American Bell, the largest of the local Bell companies, was headquartered.

While the telecommunications business was growing and the Bell companies were gaining market share, a growing public sentiment against monopolistic practices was emerging. Two new concepts emerged through Supreme Court decisions in the latter part of the nineteenth century. *Munn v. Illinois* (1877) established the concept of public utilities through a case dealing with electric utilities. In 1898, the *Smyth v. Ames* decision established the right of public utilities to a fair return, determined by government regulators, thereby creating the concept of a *regulated* public utility.

In the early part of the twentieth century, American Telephone and Telegraph began a program of acquisition to extend its market reach and thwart rivals' growth. Among the acquisitions of AT&T was a controlling interest in Western Union in 1909. This gave AT&T control over both telegraph and telephone communications. This situation drew the attention of policy makers at the state and federal level. At the state level, where legislators were more responsive to the smaller independent telephone companies, legislation requiring the AT&T companies to allow independent companies to interconnect was passed in 34 states between 1909 and 1934. At the federal level, The Mann-Elkins Act (1910) extended the authority of the Interstate Commerce Commission to include regulation of rates and services provided by interstate telephone and telegraph carriers. (This would later help explain the structure of the Communications Act of 1934.)

The "Kingsbury Commitment" in 1913 was AT&T's reaction to the mounting regulatory pressure. Enforcement of antitrust laws was an important campaign issue in 1912. Following up on the new authority granted to the government by Mann-Elkins, George W. Wickersham, Attorney General under incumbent president William H. Taft, requested that AT&T curtail its growth through mergers and acquisitions until a government investigation could be completed. The investigation was completed and Wickersham's report, which concluded that AT&T was not in violation of antitrust laws, was issued shortly before the elections. Public reaction helped candidate Woodrow Wilson win the election. The Wilson administration picked up where Taft had left off; with two new pieces of legislation, the Clayton Act and the Federal Trade Commission Act (which would establish the Federal Trade Commission) moving forward, AT&T President Theodore Vail decided to pre-empt federal government action and offer a solution that AT&T could live with. This offer came in the form of a letter from N.C. Kingsbury, an AT&T vice president, to Attorney General James C. McReynolds. This letter, later to become known as the "Kingsbury Commitment," led to the separation of Western Union from AT&T (although AT&T maintained a position in the private telegraph market.) More significantly, the Commitment contained provisions to allow the independent telephone companies to connect to AT&T lines for toll (long distance) service, and divided the telephone market along existing lines, with AT&T assured of a virtual monopoly in the markets it served at the time (which were primarily the more profitable urban centers) while the less profitable rural areas were to be left to the independents. AT&T agreed not to acquire any of the independents without approval of the Interstate Commerce Commission. Agreement with the letter signaled the government's acceptance of the AT&T's concept of "Universal Service," a concept first publicly described by Vail in 1907. According to this view, Universal Service required a single national system, operating under a single policy. The Commitment established AT&T as the system provider and policy maker: the independents would need to adhere to AT&T's rules in order to be able to provide long-distance service to their customers.

The Communications Act of 1934 was the next major regulatory step. Patterned in part after the 1887 Interstate Commerce Act, this Act created the Federal Communications Commission, which was itself modeled after the Interstate Commerce Commission (and which had been given the charter to regulate the telephone and telegraph industry under the Mann-Elkins Act of 1910.). The 1934 Act established into law the concept of universal service, stating it's purpose to be, "To make available, so far as possible, to all the people of the United States a rapid, efficient, Nation-Wide, and world-wide wire and radio-communication service with adequate facilities at reasonable charges." The 1934 Act in many ways served to consolidate various pieces of legislation, including the Radio Act of 1927, and much of the Act dealt with the broadcast radio industry. The 1934 Act also clearly defined the role of federal versus state regulators, with the FCC responsible for interstate rates and also for regulating interconnections between

AT&T and smaller companies, as well as regulation of radio, while state public utility commissions had authority to regulate communications services within their individual borders. The 1934 Act was based on the technologies of the time; data communication was never a consideration.

The Growth of Data Communications

Voice communications remained the primary consideration of the FCC and the telephone companies for the next two decades. In 1958, Bell Telephone began to offer its Data Phone service for transmission of data over regular telephone lines. By the late 1960s, business demand for data services had grown significantly, and the FCC began its Computer Inquiry I investigation. Computer I resulted in the differentiation of computer processing from communications services. Computers used for communications signal switching applications would fall under the terms of common carrier offerings in Title II of the 1934 Act, while data processing functions including data sorting, storing, calculating, etc., provided to users over the telephone system would fall under the ancillary services provisions of Title I. The Computer Inquiry I decision in 1971 established rules allowing large phone companies to engage in data processing services only if they set up separate corporate entities to preclude burdening regulated operations with costs from non-regulated competitive operations. AT&T and the Bell subsidiaries were already excluded from data processing services under the terms of a consent decree in 1956. In subsequent Computer Inquiry II and III rulings, and in rulings subsequent to the Telecommunications Act of 1996, the FCC would refine the definitions of basic and enhanced services, eventually evolving to "telecommunications" and "information service," choosing to restrict regulation to the basic (telecommunications) offerings and exempting the information services, which would include Internet service providers, from regulation. Data transmission over telephone lines would thus be regulated because the basic operations or protocol conversions needed to get the digital data over the wires did not constitute the type of manipulation or storage that altered the content of the data. Without special corporate separation and FCC approvals, the Bells remained shut out of the information service markets, although they saw their traffic continue to grow as the ratio of data to voice traffic on local exchanges, which provided access to the Internet service providers, grew with the advent of personal computers.

Until the late twentieth century, home computer users and many small businesses were satisfied with the data rates available over dial-up connections to an Internet Service Provider. The Universal Service concept ensured that virtually all Americans had the telephone service that this method required. As we will see later, new services available on the Internet drove a demand for faster access. Despite its ubiquity, the telephone industry is not the dominant provider of broadband Internet connection to the home market; that distinction belongs to the cable television companies.

Cable television began as Community Antenna Television (CATV) in 1948, a result of post-war entrepreneurship and the desire of people living in rural areas to access television broadcasts. By the 1960s, antenna and amplifier technology had advanced significantly; CATV systems were using coaxial cable to provide the broad bandwidth signals needed to carry many television channels simultaneously. The idea of digital signals sharing this bandwidth was still decades away.

By 1965, the CATV industry had grown large enough to attract the attention of the FCC, which issued regulations for the use of signal transmissions using microwaves. A U.S. Supreme Court ruling in 1998 (*United States v. Southwestern Cable Co.*, 392 U.S. 157) verified the authority of the FCC over the cable industry. Subsequent FCC regulations regarding cable more closely resembled those applied to the television broadcast industry than to the telecommunications industry, focusing on issues such as franchise standards, program non-duplication, and technical standards. Cable television, as CATV came to be known, continued to advance technically and geographically. Fiber optic lines began to be employed in the early 1980s, and by the 1990s hybrid fiber optic and coaxial systems were becoming the norm, with fiber optics used for digital transmission of signals to neighborhood "nodes" from which coaxial cable would then take over for the connection to homes. The cable television infrastructure thus had a significant bandwidth advantage over telephone, and at least some parts of the cable infrastructure were designed to handle digital signals.

The Cable Communications Policy Act of 1984 (the Cable Act) amended the Communications Act of 1934 to clarify jurisdictional authority among federal, state, and local governments over cable television systems, and established policies in the areas of ownership, subscriber rates and privacy, obscenity, unauthorized reception of services, and other areas. Only a single cable operator served most communities, and cost increases were far outpacing inflation in the 1980s and early 1990s. In response to consumer complaints, Congress stepped up regulation of the cable industry. The Cable Television Consumer Protection and Competition Act of 1992 was intended to encourage cable operators to expand their capacity and program offerings while protecting consumers against non-competitive price increases. By 1996, the mood in Congress had changed; increasing competition and decreasing regulation was the theme of the day. (The Telecommunications Act of 1996 will be examined in more detail shortly.) In a serendipitous convergence of technology and legislature, 1996 saw the roll-out of Internet access via cable modems, which, because of the inherent infrastructure advantage, could offer data rates up to 100 times faster than were available over telephone company dial-up lines.

Computers: Complement, Competition, or Convergence?

While the computing industry can trace its roots back to the inventions of Jacquard and Babbage, slightly pre-dating Morse's telegraph, significant advances in computers were not made until electronic devices, first vacuum tubes, later transistors, then integrated circuits became available. AT&T was well established as a monopoly before IBM built its first electronic computer. Because computers were designed to be "stand-alone" devices that digested reams of punched cards and produced printed forms, there was not the infrastructure investment in rights-of-way, wires, poles, and switches that the telecommunications business required. Although substantial research and development investment was needed, the computer business was more open to competition, and did not fall under the category of regulated public utility. This does not mean that the government has ignored the computer industry, as both IBM and more recently Microsoft have had to deal with antitrust charges. Nevertheless, the degree of competition in the computer and software field has far exceeded that in the telecommunications field. A business or home computer buyer can choose from competing products, unlike the telecommunications business where (at least on the local level) there is generally only one carrier land-line available. (Some may argue that Microsoft is so dominant in the operating system field with its Windows products that there is no real choice; this topic is addressed in a later chapter.)

Personal computers came on the scene in the mid 1970s, initially as a hobbyist phenomenon but soon finding a place in businesses and eventually in homes and schools. Software programs for text editing (later to become known as word processing, then expanding into *office automation*) were among the earliest applications, not unexpected since a typewriter-like keyboard was used for data entry. Other early applications, such as spreadsheets, allowed for manipulation of numbers, a requirement for business and financial use. The earliest personal computers also had game programs, establishing computers at least in part as entertainment devices.

Two significant advancements were made that increased the demand for personal computers: the Internet and the web browser. The Internet provided means for computer users to communicate, initially for sharing research data, later for sharing anything from recipes to music video files. The Internet began as the Advanced Research Projects Administration Network (ARPANET), a Department of Defense (DOD) initiative in the 1960s. In the 1980s, the DOD transitioned its activity to a separate Defense Data Network, but by that time a separate NSFNET, funded by the National Science Foundation (NSF), was in operation. In 1992, the NSF began to phase out its support for what had come to be called the Internet. Today, the high-bandwidth "backbone" of the Internet is operated by private companies including MCI, Sprint, and GTE. The backbone carriers interconnect at Network Access Points, which allow network traffic to flow without regard to which provider operates a particular backbone segment. The backbone operators sell access to Internet Service Providers (ISPs), who in turn sell the Internet access to home and business users. They in turn rely on the connection provided by their local telephone or cable television service to connect to the ISP.

The web browser simplified the user interface to the Internet and to the computer itself, taking some of the technical mystique away from the box and making it more user-friendly. Web browsers provided the ability to use graphic images, in addition to text, and simplified existing tools such as email, file transfer protocol, and Usenet news services. This drew in a new wave of Internet users.

The role of the Internet as a communications media was becoming clear. In 1995, the Federal Networking Council[1] provided an official definition of the Internet:

> Internet refers to the global information system that—(i) is logi-
> cally linked together by a globally unique address space based on
> the Internet Protocol (IP) or its subsequent extensions/follow-ons;
> (ii) is able to support communications using the Transmission
> Control Protocol/Internet Protocol (TCP/IP) suite or its subse-
> quent extensions/follow-ons, and/or other IP-compatible proto-
> cols; and (iii) provides, uses or makes accessible, either publicly
> or privately, high level services layered on the communications
> and related infrastructure described herein.[2]

A simpler definition was codified in the Telecommunications Act of 1996, which defined the Internet as "the international computer network of both Federal and non-Federal interoperable packet switched data networks."

By 2002, the volume of data traffic had exceeded the volume of voice traffic on the telephone systems in many areas. As the number of network users grew, a phenomenon known as Metcalf's Law[3] had come into play: the value of a network increases by the square of the number of nodes on the network. Businesses were attracted to a new way of reaching their customers, and businesses such as eBay and Amazon.com created new business methods to take advantage of the new medium. Banks, schools, and governments all jumped on the Internet as a way to

---

[1.] The Federal Networking Council (FNC) was chartered by the Committee on Computing, Communications, and Information (CCCI) of the National Science and Technology Council, which was formed by the Office of Science and Technology Policy during the Clinton administration. The FNC's functions have been transitioned to the Large Scale Networking Office of the National Coordination Office for Information Technology Research and Development, which replaced the CCCI under the George W. Bush administration.

[2.] Federal Networking Council Resolution, October 24, 1995. Available on line at *http://www.itrd.gov/fnc/Internet_res.html*.

[3.] Robert Metcalf is the inventor of the Ethernet and founder of 3Com Corporation.

expand services without adding personnel or opening new offices. Combined with Moore's Law,[1] which predicted regular improvements in computing power at the same or lower cost, Internet access became both desirable and increasingly affordable. Of course, some segments of the population and geographic regions have less access than others. This situation has come to be known as the Digital Divide, and is addressed in Chapter 9.

The Telecommunications Act of 1996

While the Internet was rapidly developing, the Telecommunications Act of 1934 remained the law of the land. That act had no provisions for digital data transmission by the regulated telephone companies, Internet Service Providers, or cable television systems. The Telecommunications Act of 1996, an amendment to the 1934 Act, was the government's move to update regulatory policies in recognition of changes in technologies and in the marketplace. Title I of the Act addresses telecommunications services, with the goal of increasing competition. Title I also attempts to codify Universal Service, the provision of basic telecommunications capabilities for all Americans. The Act recognizes the differences between the technology in the older, established voice communications marketplace and newer technologies applied in the faster-growing cable television and data communications field. The Act also addresses regulations that require sharing access to services or infrastructure ("unbundling") to promote efficient deployment of essential advanced infrastructure technologies.[2] Title III deals with cable television, deregulating all but basic cable television services, opening the door for the cable industry to rapidly become the dominant provider of broadband data services to home users, and allowing the cable operators to expand into voice communications. At the same time, the Act allowed for the entry of incumbent telephone companies into the cable TV market. Other sections of the Act deal with regulatory reform within the FCC, changes in regulations of the broadcast radio and television industry, control of obscenity and violence in programming, and encouragement of socially desirable goals.

Not long after the 1996 Act became law, industry groups began lobbying for changes that would affect telecommunications as related to the information technology community. The Bells had (and still have) a virtual monopoly on local

---

[1.]   Moore's Law refers to an observation made by Gordon Moore, co-founder of Intel Corporation, in 1965 that the number of transistors that could be fabricated per square inch on integrated circuits had doubled annually since the integrated circuit had been invented.

[2.]   The concept of unbundling is intended to promote competition by allowing alternative service providers to lease access to the existing infrastructure; otherwise, competitors would be blocked unless they ran their own wires to a customer's home or office.

telephone service by virtue of their control over the "last mile" of the wire that enters consumers' homes. (Technology, in the form of cable, wireless, and satellite, is changing this picture, as will be seen later.) The Act had clearly identified Internet access as a component of *Universal Service*. However, since the Act, extension of broadband Internet access through the incumbent local exchange carriers (ILECs), primarily the Bell companies, had been progressing slowly. In an effort to accelerate the introduction of broadband Internet access to homes, Rep. W. J. "Billy" Tauzin, R–LA, Chairman of the House Energy and Commerce Committee, along with Rep. John Dingell, D–MI, ranking Democrat on the committee, proposed the Internet Freedom and Broadband Deployment Act. The underlying logic was that the local exchange carriers were reluctant to invest in the improvements to their systems that would allow more broadband access because the new services such investment allow would make the local markets more attractive to competition, who could enter the market without the same capital investment and therefore significantly lower risk. The bill also attempted to level the playing field between the local exchange carriers and the cable television companies. Cable providers are the dominant broadband Internet service provider to the consumer market and yet this part of their service is not regulated (the cable television industry was "deregulated" except for basic cable television service as one result of the Telecommunications Act of 1996). One way the bill would accomplish this leveling was to remove the authority of state and local regulatory agencies over the Bells, allowing them the freedom enjoyed by the cable companies.

The Tauzen-Dingell Broadband Deployment Act, officially HR 1542, the Internet Freedom and Broadband Deployment Act of 2001, would have modified certain provisions of the 1996 Act to make conditions more favorable to the local Bell companies. This bill would allow the local Bells to offer high-speed Internet services over long-distance lines without first requiring them to open up their local service areas to competition. This would modify Title I of the Act which tied the Bell's ability to offer long distance voice or data service to their opening up of local markets to increased competition. The Tauzin-Dingell Bill was passed by the House of Representatives on February 27, 2002. When it was brought to the Senate, it immediately ran into opposition from Senator Hollings (D–SC), then Chairman of the Senate Commerce Committee The Act died in the committee, with the Senate preferring to let the Federal Communications Commission refine the regulatory rules needed to bring about incremental changes in the current situation.

Although the Tauzin-Dingle Act did not become law, rule-making action by the FCC in early 2003 did provide the Bells much of what they sought. Since the Act of 1996, the Bells have been required to allow other companies unbundled access to their local lines. Intense lobbying efforts on the part of the Bells, led by SBC, to change these rules have been countered by equally intense efforts by the long-distance carriers and other interest groups. The rules have allowed the traditional long distance carriers to move into local markets, to the extent that they now have more than 10 million customers for combined local and long distance packages. This competition has resulted in competitive price-cutting in some markets. On February 20, 2003, the FCC voted to adopt new rules, which, "provides incentives for carriers to invest in broadband network facilities, brings the benefits of competitive alternatives to all consumers, and provides for a significant state role in implementing these rules." This decision was marked by dissent within the FCC, with four commissioners dissenting with various parts of the decision, and only one agreeing in whole. The two main points of this decision were to return significant authority to the states in regulating local telecommunications services, and to eliminate some "unbundling" requirements. The Commission ruled that incumbents do not need to unbundle the high-frequency portion of the copper loop circuits that are used to provide high-speed digital subscriber line (DSL) service. Currently, approximately 40 percent of DSL service is provided by companies that "share" the incumbents' copper to the home lines. This step will effectively reduce competition in the broadband market and increase process to consumers. The impact is expected to be most noticeable in rural areas where the local exchange carrier (LEC) has not yet decided to offer broadband service due to the low population density. The inability to share lines would force these local Internet Service Providers (ISPs) out of business unless they are able to lease dedicated lines at significantly higher prices than they currently pay for shared service. FCC Chairman Michael Powell dissented strongly with the majority decision on this point. Powell's position was that the DSL service uses existing copper lines, therefore sharing of those lines does not factor into decisions on investment in new fiber optic infrastructure. The Commission also ruled that where the LECs invest in new fiber optic installation, they need only "share" low-bandwidth voice-grade capacity on those lines. These steps were intended to encourage the incumbents to invest in infrastructure improvements by removing the need to share these improvements with competitors.

The decision to allow increased state regulation is intended to give state public utility commissions the deciding voice in local "unbundling" decisions. This step allows local competitive and economic factors to be considered in these decisions, rather than national circumstances. This action opens the possibility that the LECs will need to comply with disparate regulations across the 50 states and the District of Columbia. On the other hand, it could result in decisions that result in lower prices and increased competition at the local level.

Reaction to the FCC decision was not long in coming. The Bells, although apparently the winners in that they achieved the position they sought in not having to bundle their high-speed DSL services, complained that the decision did not go far enough, and threatened to seek remediation in the courts. Their complaint is that they still are required to share local voice-grade lines, and that the regulated rates they are allowed to charge are below their cost. That means, they claim, that their profits from other services are underwriting the cost of providing local voice competition, and draining funds that could otherwise be used for investment.

Stakeholders

Industry and Consumer Groups

The Tauzen-Dingle Act unleashed a level of lobbying and media campaigns that brought the behind-the-scenes activity of the legislature and the special interest groups into the public spotlight. An examination of the players involved in that case, and in subsequent regulatory decisions, provides a good example of the current stakeholder line-up.

Two camps were formed around the Tauzen-Dingle Act. Those for the bill included the Bells, the Communications Workers of America (the union that represents the majority of the Bell's organized labor force), and the United States Telecom Association (a group that represents over 1,200 companies, mostly independent or competitive local exchange companies, as well as the Bells themselves). Opposing the bill were the major long-distance carriers AT&T and MCI Worldcom, the cable television industry, the Competitive Telecommunications Association, The National Association of Regulatory Utility Commissioners, the Consumer Federation of America, and an organization called Voices for Choices, funded by the long distance carriers, which attempted to rally support against the bill through television and newspaper advertisements.

The Bells supported this bill on the basis that it would provide incentive for investment in new technology and infrastructure improvements needed to bring broadband Internet access to rural areas and areas not served by current DSL capability. Without the protection of this bill, they could not expect a reasonable return on their investments. Those in support of the bill argue that they are faced with regulatory burdens that are not imposed on competitive telecommunications delivery systems (particularly cable TV and satellite), and that they should be afforded some protection of their markets in return for the regulatory burdens imposed on them.

The opposition to the bill focused on claims that it would continue the current monopoly situation in local telephone markets, effectively removing the possibility of competition among providers and choice for consumers. Further, the bill would add a new service offering, broadband Internet, to the Bells' local monopoly. These arguments are based on the logic that the Bells won't open up local markets unless forced (or coerced) to do so; removing the restrictions that tie their ability to move into long distance voice and data to the opening of local markets effectively neutralizes any regulatory leverage to cause this to happen. The cable television industry opposed the bill because they currently provide broadband Internet access to 70 percent of those who have such service and see the Bells move into this market as a significant threat. State and local regulators saw the bill weakening their authority. Consumer groups opposed the bill on the grounds that it weakened the increased-competition provisions of the 1996 Act.

Government

The House Committee on Energy and Commerce has jurisdiction over legislation dealing with telecommunications within the House of Representatives. This committee, the oldest legislative standing committee in the U.S. House of Representatives, traces its origin back to the Committee on Commerce and Manufactures, which was created on December 14, 1795. Under the Committee on Energy and Commerce, the Subcommittee on Telecommunications and the Internet, considers legislation dealing with interstate and foreign telecommunications, by all modes, including broadcast, radio, wire, microwave, satellite, or other mode.

In the United States Senate, legislation dealing with interstate commerce, communications, and science, engineering, and technology research and development and policy is referred to the Committee on Commerce, Science, and Transportation. The Communications Subcommittee has jurisdiction over legislation dealing with telecommunications and the FCC.

Federal Communications Commission (FCC)

The FCC is an independent government agency created by Congress with the Communications Act of 1934. Its mission is to regulate interstate and international communications by radio, television, wire, satellite, and cable, within the 50 states, the District of Columbia, and U.S. possessions. There are five commissioners, appointed by the president and confirmed by the Senate. No more than three commissioners may be from the same political party. Each commissioner serves a five-year term, and they are prohibited from having any financial interests in businesses regulated by the FCC.

There are six bureaus and ten offices in the FCC (see Figure 2–1). Of particular interest to the topic of telecommunications and computing are:

- The Consumer and Governmental Affairs Bureau (CGB), which is responsible for informing consumers about telecommunications goods and services and seeking their input to the Commission on subjects of interest to consumers. CGB also coordinates policy efforts with the telecommunications industry and with other governmental agencies.

- The Wireline Competition Bureau is responsible for rules and policies concerning telephone companies that provide interstate, and under certain circumstances intrastate, telecommunications services to the public through the use of wire-based transmission facilities. This includes the Internet backbone services as well as local voice and data communications.

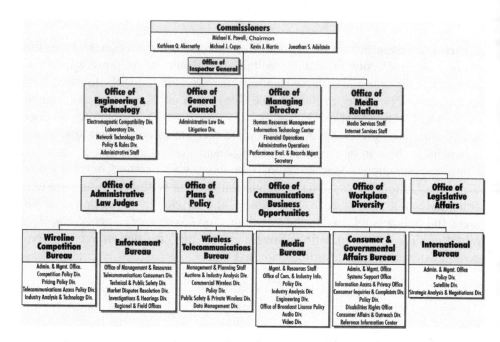

Figure 2–1   Federal Communications Commission Organization (March, 2003).

- The Office of Engineering and Technology provides expert advice on technical issues before the Commission and allocates spectrum for non-governmental wireless communications. This office is responsible for (among many other tasks) performing objective analysis of new network technologies and assessing the impact these new technologies may have on the competitive market. They also work with industry and other government offices to develop standards for telecommunications networks and interconnectivity.

Court Decisions and Regulatory (FCC) Actions

There are several milestone cases since the mid-twentieth century that are critical to an understanding of how the telecommunications industry and telecommunications policy got where it is today. Before World War II, the FCC began to investigate AT&T's monopoly of telephone equipment through its Western Electric subsidiary. No action was taken during the war, as AT&T was considered to be performing a valuable and unique service for the war effort. In 1949, the Department of Justice filed a case against AT&T, asking that the company divest its equipment-manufacturing subsidiary and license its patents to other manufacturers, in order to promote competition in the communications equipment market.

After seven years in the courts, a consent decree was reached in 1956 in which AT&T would keep Western Electric, but would license its equipment patents to others. Furthermore, AT&T (including Western Electric) would limit their business to telephone communications. This decision by AT&T would have downstream effects as it shut the door for their timely entry into the computer industry and also provided key technologies, including the transistor, to that same industry.

The Execunet decision in 1968 is another milestone case, because it opened the door for competitors to connect to AT&T's local telephone switching service. Execunet was the name used for a private-line service offered by Microwave Communications, Inc. (a company that later changed names to MCI, then MCI Worldcom, then back to MCI again) to business customers. Microwave Communications wanted to be able to offer these customers access to local calls, which required connecting to the local switches. The court decided in favor of AT&T, but Microwave Communications won on appeal. This decision increased competition in the long distance telephone market, because service offered by competitors to AT&T became more useful, and valuable (another example of Metcalf's Law) with the ability to connect to AT&T's local service.

Not long after, in 1978, the Department of Justice again filed an antitrust case against AT&T, this time alleging that profits from its regulated monopoly telephone business were being used to undercut prices and exclude competition in the telephone equipment business and that AT&T was using the regulated price structures for local service to discourage competitors from connecting to its networks. Judge Harold Greene presided over the case that would last until another consent decree was reached in 1982. That decree, technically a modification of the 1956 decree and therefore known as the "Modified Final Judgment," largely architected the telecommunications industry as we know it today. Under this agreement, AT&T would divest the local operating companies (which became known as the Baby Bells). AT&T would keep its long distance operations, the Western Electric Company, Bell Labs, and the Yellow Pages business. Long distance service would continue to be regulated by the FCC, but AT&T would be allowed to compete in some computer communications and wireless services.

Technology: Increasing Capability and Competition

Cable, Copper, Fiber

The recent FCC decision to unbundle the Bell's DSL lines in order to spur investment, presumably in new fiber optic infrastructure, could in fact have the opposite effect, thanks to technological advances that have improved the bandwidth of the existing copper lines that serve the vast majority of American homes and businesses. DSL technology today can achieve a maximum data rate of 1.5 megabits per second, and is available only within a relatively short distance from the switching system. Experimental technologies, called VDSL for very-high-data-rate DSL) can increase that rate dramatically over distances of 3,000 to 4,000 feet. By extending a new fiber optic backbone through neighborhoods and using VDSL over existing copper wires to individual homes, data rates of up to 50 megabits per second could be possible. At that rate, consumers would have access to services like on-demand movies, enhanced video teleconferencing, more realistic interactive gaming, and virtual shopping that are only dreamed of today. Under the current FCC guidelines, however, these lines need not be shared; consumers will not have access to a choice of DSL providers.

Would consumers pay the extra cost for higher speed Internet access service? Currently, about 14.8 million households subscribe to a broadband service, with the majority (9.4 million) being through a cable television service, and most of the rest (5.4 million) being telephone DSL service. Only about 20,000 homes have direct fiber optic connections, and these are for the most part new construction homes in "wired" developments. Broadband service is not available in all areas. Cable television is generally limited to densely populated urban and suburban areas, and DSL is limited to homes and offices in close proximity to the telephone company switching centers. Broadband service typically costs $35 to $45 per month for cable modem service, whiled DSL service ranges from $50 to $70 per month, with some difference in rates based on speed. Although broadband service is currently available in most metropolitan areas and is moving rapidly to suburbs, less than one-third (about 19 million) of the 65 million households with Internet service subscribe to a broadband service. Cost is one obvious factor, perceived value is another. As broadband service becomes more widely accepted as a tool for entertainment, communications, and education, as well as home security, telecommuting, and shopping, the perceived value relative to current telephone and/or cable TV bills, not to mention commuting expenses such as gasoline and parking, will increase. The price reduction, though, will depend on competition, which is slow in coming. Since deregulation, cable television rates (other than basic access) and Internet access via cable have risen steadily, primarily due to the absence of real competition.

Satellite Internet Service

An alternative to the DSL service offered by the phone companies or cable modem service available in some areas is Internet service via satellite. This option is available anywhere, as the ads claim, where a clear view of the Southern sky is available, in order to have a clear line-of-sight with geo-synchronous communications satellites. Satellite service is therefore available almost anywhere, even remote rural areas that are outside of the DSL limits or not served by cable.

Performance-wise, satellite is a distinct improvement over dial-up telephone Internet access, but is significantly slower that either DSL or cable modem service, and significantly more expensive. Down-link data rates of 400 Kbps can be achieved with satellite, compared to 1.5 Mbps with cable or DSL. Uplink rates are slower, in the 60 Kbps range for satellite, compared to 128 Kbps for DSL and 384 Kbps for cable. The cost of satellite service is typically higher than either cable or DSL ($69 v. $39 per month) and an initial investment of around $800 in equipment and installation is required. (Unlike receive-only satellite television systems that can be installed by a do-it-yourselfer, satellite Internet systems which have transmit as well as receive capability must be installed and aligned professionally.) Until performance and pricing improve, satellite Internet service will be the last resort of the well-heeled and desperate web surfer.

Wireless Internet

Advances in technology for data transmission over wireless networks are providing yet another technology challenge for the telecommunications industry and regulators. Known technically as IEEE 802.11a, (or b, or more recently, g) and known popularly as "Wi-Fi," wireless networking has evolved in a few years from a technologist's curiosity to an out-of-the-box solution available to any home or business computer. Wi-Fi technology has become a standard feature in high-end notebook computers and personal digital assistant (PDA) devices, and can be found in some cellular telephones. Industry analysts predict that Wi-Fi will become the de facto standard for home networks, used for interconnection of computing and home entertainment devices.

Wi-Fi uses low-power signals broadcast in the 2.4 GHz and 5 GHz frequency bands set aside by the FCC for unregulated consumer-product use. The relatively low power (and therefore short range) signals used, combined with a technique known as spread spectrum that broadcasts on multiple channels to overcome interference, allow Wi-Fi to share the allocated bands with cordless phones and other consumer products. Design features in the transmitting and receiving devices allow overlapping Wi-Fi networks (as might be found in an office or apartment building) to operate on a non-interfering basis, although data privacy remains a concern. Some hotels, cafes, and even college campuses are installing Wi-Fi; users with compatible equipment need only locate a "Hot Spot" where the signal is available, then they can work on their computers with the same high bandwidth available in a hard-wired office. The system does not yet allow seamless roaming from one hot spot to the next (unlike cell phones, which maintain connectivity as the user drives down a highway), but that level of sophistication may not be far off. Wi-Fi data rates surpass those of DSL or cable, but are currently very distance-limited.

How does Wi-Fi challenge the industry? Because it is wireless, Wi-Fi presents an alternative to the telephone and cable companies for broadband data connectivity. Minor adjustments in power levels or the use of directional antennas could allow Wi-Fi or its technical successor to provide an alternate signal into homes, bridging the "last mile." Furthermore, when coupled with technology to allow voice communications and interconnection with the existing telephone infrastructure, Wi-Fi could allow its users to unplug completely.

Voice over Internet Protocol

The ability to convert analog signals such as voice to digital data, and vice versa, traces back to Shannon's paper mentioned in the introduction to this chapter. When coupled with the Internet Protocol used to exchange packets of data among computers, a technology known as Voice over IP results. VoIP uses the same network, either wired or wireless, to transmit packets of digitized voice data as are used otherwise to transmit "regular" (non-voice) data; the network doesn't care what the data means because, it's all 1's and 0's. A special device called a gateway serves as the interface between the telephone in a subscriber's home or office and the data network, and at the other end between the data network and the public switched telephone network, operated by the local exchange carriers. Access to a high-speed data network is necessary, otherwise the quality of the reconstructed voice signal will be poor. For this reason, VoIP systems are finding their earliest use as replacements for the PBX systems in offices that are already wired with high-speed local area networks (LANs).

VoIP service is currently offered by some cable companies as an option to their cable modem service, as well as by some independent carriers. VoIP services are aggressively priced to compete with local and long distance telephone providers, and offer features that are unavailable from the LECs. A typical VoIP package offers unlimited local service plus 500 minutes of long distance, and also includes a variety of services, including caller identification, call waiting, call return ("*69" service) and voicemail, for a bundled price of what a LEC would charge for the local phone service alone. Some VoIP systems allow the subscriber to select the area code of his choice. For example, from a home office in Virginia, one could have a telephone number with a 210 (San Antonio, Texas) area code. This would be useful to someone trying to establish a business presence in another area.

Obviously, this type of service is not available to everyone. VoIP services are viable options only to those who have broadband service available at their home or business. The cost of the broadband access must be factored into any subscription decision, but as this is generally considered to be a necessary business expense, VoIP service appears to be an up-and-coming competitor to the LECs, especially where alternative broadband connections such as cable are available. Currently VoIP is not regulated as a telecommunication service, but is instead considered to be in the unregulated enhanced data service category. Perhaps the FCC's vision of increasing competition through deregulation, resulting in lower cost to consumers, is playing out in this case.

Computers As Communications Devices

As the VoIP example points out, the use of computers as communications terminals has evolved dramatically since the early days of email. Instant messaging, a technique pioneered by AOL, allows interactive text "chat" sessions to take place between users across a room or around the world. Software packages such as Microsoft's NetMeeting allow appropriately equipped users to engage in a real conversation, and even to view a real-time video image of the person with whom they are speaking, making the computer into a "picture-phone," at least for those with network access. As Figure 2–2 shows, email remains the most common use of the Internet today, but there are clearly a number of other high-volume activities.

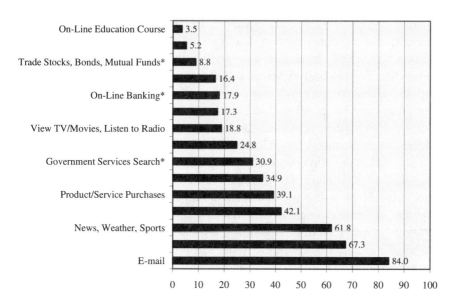

Figure 2–2   Activities of Individuals Online, 2001 as a Percentage of Internet Users, Persons Age 3 +.

(Source: *A Nation On Line: How Americans are Expanding Their Use of the Internet.* U.S. Department of Commerce, Economics and Statistics Administration, National Telecommunications and Information Administration. Washington, D.C. February, 2002. Available on line at *http://www.ntia.doc.gov/ntiahome/dn/index.html.* Figure 3-2, p. 30.)

Network-accessible computers have displaced significant volume of other forms of business and personal communications. online banking and bill-paying services have decreased the volume of paper checks, and online shopping has significantly reduced the volume of mail-order and telephone-order business. Industry experts estimate that by 2002, the public telephone system was being used more for data traffic, including web browsing and email, than for traditional voice calls.

The government has had little involvement in regulating and standard setting with regard to computers as communications devices. Email, instant messaging, and even video-conferencing on the Internet are not subject to tariffs beyond what the user pays for connecting his/her computer to the network, typically a local phone call and a fixed ISP charge. Computer hardware standards and Internet data protocols, including those for secure transmission of personal or financial data, are not regulated by the government, they are set by industry or professional groups (de jure standards) or by default when one provider becomes dominant enough to sway the market (de facto standards).

Digital Entertainment

Even low-priced computers today are equipped to handle a variety of entertainment formats, including music, video, and games, in addition to regular text. Home computers commonly have stereo speakers with sub-woofers to more realistically convey the crashes and explosions common in popular computer games; even notebook computers generally have stereo sound. Computer displays are getting larger, and have better resolution than standard television. Some monitors serve dual-duty, with built-in television tuners. Alternatively, a tuner-module can easily be added to most computers. Computers are typically equipped with CD read/write units, and DVD players are becoming more common. Students moving into college dorms no longer lug speakers and turntables; their mandatory computer provides all the audio and visual capability they need.

The computer as entertainment center marks another milestone in the convergence of computers and telecommunications. Broadband communication allows alternative media outlets to compete with broadcast radio and television. Webcasting allows anyone with enough bandwidth to set up an online equivalent to a radio station, and share commentary, news, weather, even music with the world. Listeners don't need an AM or FM receiver, only Internet access and multimedia software such as RealNetwork's RealOne Player ™ or Microsoft's Windows Media Player™. Webcasters working within a large LANs, as might be found on a college campus, find this to be a lower-cost alternative to broadcast radio: less equipment, an in-place infrastructure, and no FCC license required. Only recently did the music industry step in and demand royalty agreements for the use of recorded music. The Internet is also becoming an alternative channel for commercial television and radio broadcasts, with many stations offering live webcasts of their news programs and some special events.

The combination of broadband Internet connections and computer technology has spawned a phenomenon known as file sharing. Using software that allows computer users to easily search other user's files for a particular music title or artist, file sharing has caught the attention of the music and motion picture industry, which see it as illegal duplication of copyrighted work and a loss of royalty income. Napster was the first file sharing system to gain widespread attention, and it was eventually shut down. Other services, KaZaA for example, have since sprung up, more sophisticated both technically and legally (operating internationally, behind an array of shell corporations). File sharing originally was a concern only to the music industry, but technological advances in the form of Broadband Internet, digital video cameras, digital video disks (DVDs), and low-cost multi-gigabyte disk drives have opened the field to motion picture piracy as well. The copyright holders, represented by their trade associations, especially the Recording Industry Association of America (RIAA), are lobbying Congress for increased protection and are undertaking legal actions against the so-called pirates. The Digital Millennium Copyright Act (DMCA) is one result of these efforts. (See Chapter 7 for a more in-depth look at these issues.). At the same time, new products and services are emerging to provide a legal means to purchase copyright-protected work over the Internet. Recently, Apple Computer began a music-download service for a modest fee, providing a legal alternative to KaZaA and Napster, and AOL is planning a similar offering. Software giant Microsoft has also entered the arena, signing an agreement allowing AOL to license its anti-piracy and digital rights management software. This agreement could lead to mechanisms that the music and motion picture industries need to profitably offer their products for distribution over the Internet.

Policy Issues of Convergence

As the previous sections have demonstrated, computers, the Internet, and the communications, business, education, and entertainment services that they synergistically allow are becoming a part of the lifestyle of many Americans. Many others are being left out, unable to share in the benefits that basic Internet, and increasingly, broadband Internet, access requires. There are a number of reasons for this, including the cost of the service (subscribers unable or unwilling to pay for the service); technical and geographic limits (many homes are not currently served by DSL or cable service); and the cost of the computer equipment itself. Chapter 9 provides a closer look at these factors. From a policy perspective, access to the Internet needs to be compared to the goals of access to telephone voice service that has been institutionalized under the name Universal Service.

Those goals, addressed in the 1934 Act and again in the 1996 Act, can be briefly stated as providing efficient and affordable telephone service to all Americans, including those in rural areas. Following the passage of the Telecommunications Act of 1996, the FCC reported back to Congress that the Universal Service provisions of the Act would be interpreted to allow special Internet access rates for schools, libraries, and hospitals, in order to support education and research missions. The FCC further determined that the inclusion of basic telephone service under the Universal Service policy was sufficient to meet the Internet access needs of citizens, reasoning that voice-grade access to the telephone system is adequate to provide dial-up connection to an Internet Service Provider. While it is true that the connection can be made, the limitation imposed by the dial-up data rate excludes such subscribers from many of the benefits of the digital age. Furthermore, the ISPs themselves are not regulated, so the dial-up method only goes half way to meeting the Universal Service objectives. The telephone service to connect to an ISP is covered, but the access to the Internet provided by the ISP is not. When the idea of Universal Service was put forth by Vale in 1907, and later codified in the 1934 Act, what we now call voice-grade telephone service was state of the art communications; this is no longer the case. Relying on the open market to provide high-speed Internet access in an unregulated environment is likely to result in competitive service only in densely populated urban and suburban areas, and even those areas are frequently limited to a single provider based on the technical limits of DSL and the infrastructure costs of updating cable systems. Federal telecommunications policy needs to be revised to consider broadband Internet access under the Universal Service umbrella.

Conclusion

Communications technology has evolved far from the days of the Morse telegraph, and government oversight of that business has come full circle with the current attitude of deregulation. The computer has evolved far beyond the limited-capability number-cruncher envisioned by Thomas Watson in the early days of the industry. While there remains a place in government, business, and science for powerful supercomputers, for example in census data processing, weather forecasting, corporate accounting and billing, or space exploration, computers today are increasingly fulfilling Scott McNealy's vision. PCs have become entertainment appliances and communications terminals. While they are widely used for "office automation" functions that have replaced typewriters and calculators, they are used more and more for communications with other computer users, or for functions such as online bill paying, online shopping, researching remote database services, or other functions that require a network connection. Our computing experience, as we have come to know and enjoy it, would not exist without our telecommunications system.

The issues that we must come to grips with revolve around providing economic incentives for technology and infrastructure investments, ensuring that services are fairly priced, and providing an adequate level of service to all. The Telecommunications Act of 1996 clearly extended the notion of Universal Service to include Internet access, but no policy yet exists to make access to the current state of the art in Internet service as truly universal as basic telephone service has become. For the time being, user fees are being used to subsidize services at schools and libraries but there is no national policy (other than to allow the unregulated market to fill the gap) to provide reliable and affordable high-speed access to all Americans. Eventually, technology will develop competitive alternatives. The distance limitations on DSL are being stretched, and cable systems are upgrading. Some day even the electric power lines may serve as a conduit to the Internet. New wireless technologies may eventually provide another alternative to the "last mile" connection. However, without a national policy, and some regulation or incentive, there is no focus to the research that is needed to accelerate these improvements, and there may be significant wasted expense in research and deployment of new technologies and competing standards until a clear winner emerges in the marketplace. High-speed Internet access is becoming as important in the twenty-first century as basic telephone service was in the twentieth century. Unless it is fully embraced under the concept of Universal Service, there will continue to be significant segments of the population that are un-served, or under-served, with regard to Internet access.

Glossary

Broadband   Originally used to define a wide range of the frequency spectrum over which multiple signals could be broadcast simultaneously, broadband has entered the computer vocabulary as an adjective describing data rates that are higher than "*narrowband*," or anything higher than 56 kbps. As defined by the FCC, broadband service must be capable of at least 200 kbps both upstream and downstream. Broadband providers today advertise rates of up to 1.5 million bits per second (mbps) *downstream*, and 384 kbps *upstream*.

Cable Modem   A modulator-demodulator devise that allows the data signals carried over cable television systems' coaxial cables to be sent from and received by a subscriber's computer.

Cable Television   A method of sending television broadcasts to subscribers' homes over coaxial cable, rather than broadcasting "over the air." The coaxial cable used allows higher frequency signals to be carried than can be achieved over the copper wire pairs used in the telephone system. This means that the cable system has more bandwidth into a home than does the telephone system (see *broadband*).

CATV   Community Antenna Television. The early name given to Cable Television.

Digital Subscriber Line (DSL)   A technology that allows high-speed data to be sent on the copper wire infrastructure originally intended for voice service. Asymmetric DSL is a variant that allows high-speed data and voice to use the same lines, but the downstream data rate far exceeds the upstream data rate (hence the term asymmetric). Even though the acronym for this, ADSL, is more accurate, the short form DSL is more commonly used.

Downstream   Data communications from the Internet Service Provider to a user's computer.

FCC   Federal Communications Commission.

Fiber to the home   This term describes the use of optical fiber instead of, or in addition to, copper wires and coaxial cable to provide telephone, television, and Internet access to a home. Fiber to the home is being installed in some new developments, but is not yet being widely retrofitted due to the cost involved and the existing infrastructure investment.

Fiber Optics   Thin strands of glass fiber that carry data signals as short bursts of light instead of using electrical signals over copper wires. Data rates far in excess of those achievable with copper wire pairs or coaxial cable are possible.

IEEE   Institute of Electrical and Electronics Engineers.

ISP   Internet Service Provider.

Kbps   Kilo (Thousand) bits per second.

LAN   Local Area Network.

Last Mile   This term refers to the cable or wire lines that run from a telephone or cable television service provider's switching system or distribution point to a subscriber's home or office. The actual distance may be more or less than a mile. The last mile is important because it incurs a significant installation and maintenance expense, and is generally owned by an incumbent local telephone company. In areas served by cable television system, an alternative "last mile" connection is provided by the regional cable television company.

LEC   Local Exchange Carrier. A telephone company providing local (as opposed to long-distance) telephone service.

ILEC   Incumbent Local Exchange Carrier. An LEC that was in operation prior to the adoption of the Telecommunications Act of 1996.

Mbps   Mega (Million) bits per second.

Narrowband   When used in computer communications, narrowband is generally considered to mean the range of data rates that are achievable over a standard voice-grade telephone line. With current modem technology, that means an upper limit of 56 thousand bits per second (kbps).

PBX   Private Branch Exchange. A private telephone network commonly used within a business.

Upstream   Data communications from a user's computer to the Internet Service Provider.

VoIP   Voice over Internet Protocol. A means of sending and receiving voice signals over data networks.

Further Reading

Aufderheide, Patricia, *Communications Policy and the Public Interest*, New York: Guilford Press, 1999.

Baumgartner, Frank (ed.), *Policy Dynamics*, Chicago: University of Chicago Press, 2002.

Bolter, Walter, W. McConnaughey, Fred J Kelsey, *et al.*, *Telecommunications Policy for the 1990's and Beyond*, Armonk, NY: M.E. Sharpe, 1990.

Compaine, Benjamin and Greenstein, Shane (ed.), *Communications Policy in Transition: The Internet and Beyond*, Cambridge, MA: MIT Press, 2001.

Esbin, Barbara, "Internet Over Cable: Defining the Future in Terms of the Past," OPP Working Papers Series, Federal Communications Commission, Office of Plans and Policy, Washington, D.C., 1998.

Frieden, Rob, *Managing Internet-Driven Change in International Telecommunications*, Boston: Artech House, 2001.

Neuman, W. Russell *et al.*, *The Gordian Knot: Political Gridlock on the Information Highway*, Cambridge, MA: MIT Press, 1998.

Olfus, Dick W. III, *The Making of Telecommunications Policy*, Boulder, CO: Lynne Rienner Publishers, 1999.

Shaw, James, Telecommunications *Deregulation and the Information Economy*, Second Edition, Boston: Artech, 2001.

A Nation On Line: How Americans are Expanding Their Use of the Internet. U.S. Department of Commerce, Economics and Statistics Administration, National Telecommunications and Information Administration. Washington, D.C. February, 2002. Available on line at *http://www.ntia.doc.gov/ntiahome/dn/ index.html*

Communications Act of 1934 as Amended by the Telecommunications Act of 1996, Washington, D.C.: U.S. Government Printing Office, 1996.

Chapter 3

Internet Governance

William Aspray

"'Spam' is a four-letter word…"

—San Antonio Express-news, October 20, 2002

T he Internet presents an interesting and important set of governance challenges. The pell-mell growth and the rising influence of business interests on the operation of the Internet have strained the self-regulation model that cybercitizens created. The Internet is too important to national security and economic well-being for the U.S. government to cede authority over it, but the authority of the United States in this sphere is uneasy both because of the libertarian philosophy of Internet users and the jurisdictional issues involved in ruling a cyberdomain that respects no national boundaries.

This chapter investigates five issues related to the politics of governing the Internet. The first section discusses the authority by which the U.S. government can rule the Internet and some of the problems with the current governance model. The next two sections discuss attempted political solutions to two problems on the Internet, viruses and spam, that through scale effects have become critical barriers to Internet access for everyone. Not every group has equal access to the Internet. Chapter 9 describes the problems of access for poor, rural, and minority communities. In this chapter we consider problems of access for the disabled, who represent one-fifth of the U.S. population. The final section discusses an intellectual property issue, cybersquatting and trademark infringement, that affects the ability of businesses and individuals to retain possession of certain domain names.

Political Authority to Run the Internet

The Advanced Research Projects Agency (ARPA) was formed in 1957 as a response to the Sputnik crisis, as a way to consolidate advanced research for the U.S. military and avoid military service research rivalries that were inefficient. ARPA (later known as DARPA) began to award research contracts in computing in 1963. After early successes in timesharing, it pioneered research in the area of computer networking in the late 1960s. Contracts were issued in 1969 to begin building the ARPANET, so that DARPA contractors could more easily share data, information, files, and computing resources at remote sites. The National Science Foundation (NSF) established a network for computer science researchers (CSNET) in 1981, and in 1983 a gateway was established between ARPANET, CSNET, and a number of other military networks. Eventually, other networks were connected, and the Internet was born.

DARPA, and to a lesser extent NSF, paid for most of the research, infrastructure, and management of the Internet during these years that the net was used primarily by the defense and scientific communities. Many of the technical decisions were made by volunteer groups of researchers, such as the Internet Engineering Task Force, or by individual scientists working under contract to the federal government, but the ultimate authority resided with the U.S. government. The nature of the Internet changed entirely in the 1990s, with the introduction of the World Wide Web, the growth of private users who quickly outnumbered users from the military and research communities, commercialization, and internationalization. By the late 1990s there was an urgent need to provide better governance structures for the Internet to handle the new set of interests and problems that had arisen.

Under government contract, researchers at the University of Wisconsin had invented a name server in 1983. It enabled Internet users to remember a name instead of the long string of digits (the Internet Protocol address) that represented the location of a particular site on the Internet. The following year, a computer scientist at the University of Southern California, Jon Postel, and one of his colleagues developed the Domain Name System as we know it today.

Only a few details about the operation of the Domain Name System need to be recounted here. A root server holds the addresses on the Internet and dynamically looks up an IP address when given a domain name. Twelve copies of this machine are located around the world (most of them in the United States), but the principal root server (or A root server as it is sometimes called) is the final authority. The work is not all done by this single root server, but instead is doled out to a group of other computers in a distributed and hierarchical fashion. Each of these other computers is responsible for a top-level domain, one of the generic domains such as .com or .edu, or for one of the country code top-level domains such as .uk for the United Kingdom. The work may be further divvied up by a top-level domain server into work for another group of computers, each one responsible for a specific second-level domain, such as ac.uk for academic institutions in the United Kingdom. However, in this hierarchical system, the final authority rests with the root server. Nobody is in a position to control communications on the Internet, such as blocking or intercepting an individual message, because the architecture of the system is designed in such a way that there is no central location through which every message passes. In fact, one cannot predict in advance the route that will be taken by the data packets comprising a message as it moves from sender to receiver across the Internet. The story is very different with respect to the domain name system. One can control who is able to be a participant on the Internet, and on which part, by controlling the root server.

The organization responsible for making policy decisions about the domain name system, such as who can use which domain name and under what circumstances new domains are introduced, came to be known as the Internet Assigned Number Authority (IANA). For two decades, Jon Postel alone served as the IANA. As the Internet grew, the job became too large for one person. A private organization, Network Solutions Inc., took responsibility for administering the domain name system, while IANA continued to set the policy. Both groups took their authority from a contract with the U.S. government.

As the Internet continued to grow and it became more economically and politically important, there was pressure to institutionalize IANA—to let it be run by organizations with well-established and transparent governance practices. Network Solutions was earning millions of dollars through the registration fees paid by individuals and organizations to use the Internet, and many other organizations wanted a piece of the action. The International Telecommunications Union, a body of the United Nations tried to exert authority over the domain name system, given the international reach of the Internet and ITU's historic role in governing the international telephone system. Governments began to be concerned about national sovereignty issues as the Internet became more important to their economies, defense, and cultural lives. Battles began to occur over cybersquatters and control over the domain names for certain well-established trademarks, yet there was no body that held the authority for these matters when they crossed national borders.

In response to these pressures and as part of his Framework for Global Electronic Commerce, in 1997 President Clinton directed the U.S. Department of Commerce to privatize the management of the domain name system in a way that increases competition and facilitates international participation in the management. The following year, Commerce published a white paper, *Management of Internet Names and Addresses*, setting out its policy.

The white paper identified four guiding principles for the privatization effort. Foremost was the need to ensure the stability of the Internet, including the development of a comprehensive security strategy. Second was a preference for employing market forces to improve competition, in order to lower costs, promote innovation, encourage diversity, and increase user choice. Third was a call for a management system that represented the wide community of users both by function and by geography. Fourth was a preference for bottom-up governance (following the model of governance that had historically been used for the Internet) instead of top-down governance structures imposed by governments.

In 1998, Commerce selected a non-profit organization located in California, the Internet Corporation for Assigned Names and Numbers (ICANN), to manage the Domain Name System. The first five years of the organization's history have been turbulent. There are many different points of view about what ICANN should be doing and how well it has met its objectives. The political issue is whether Commerce is willing to continue to vest authority in ICANN for running the domain name system. There has, in fact, been significant criticism of ICANN in Congress, and the renewal of ICANN's contract later in 2003 would be seriously in doubt if the government had other viable options.

A report in 2002 by the General Accounting Office (GAO) presented the government's scorecard on ICANN. Overall, the GAO gave ICANN strong marks on the competitiveness principle set out in the white paper and weak marks on the other three principles. In terms of increasing competition, there are now 180 registrars of domain names, whereas there was only Network Solutions in 1998. Moreover, seven new domain names (with seven different managing organizations) have been added, chosen from 44 applications. The other high marks generally for ICANN have come for the establishment of the Uniform Dispute Resolution Policy for handling alleged cases of cybersquatting (described in detail in the final section of this chapter).

ICANN missed several deadlines for improving the security of the root servers spelled out in research and development agreements with Commerce. GAO criticized ICANN's progress in regularizing and formalizing agreements with the organizations running the top-level domains. As of 2002, ICANN had only reached formal agreements with two of the 240 organizations that manage country-code domains. GAO was also critical of ICANN's progress on the other two principles: wide representation and bottom-up governance.

The most serious complaints have focused on the board of directors and the way in which board members are elected. Much to the distress of Congress, which held hearings on the matter, the original board of directors was appointed without public participation of any kind. The interim board moved quickly to seat representatives with technical expertise, but slow to seat people representing other interests. Through a set of bylaw changes, the board worked to delay and reduce user representation on the board. When elections were finally held in 2000 to elect user representatives to the board, the sitting board filled the electoral slate with people from large telecommunications companies rather than select a diverse group of users. Less than half of the Internet users who tried to be verified to vote in this board election—75,000 out of 158,000—were able to do so. Then only 34,000 people voted (less than 4,000 in the United States). This was a tiny number of ballots cast compared to the many millions of Internet users. In late 2002, rather than fix the election process, the ICANN board yet again amended its bylaws to eliminate the elected at-large seats in favor of seats representing the interests of various other committees.

The tensions over representation and control of the domain name system continue to mount. They are due in part to the competing and incompatible demands of various interest groups, and in part to the self-destructive practices by ICANN. The reform measures that ICANN has initiated in its own governance procedures seem too little, too late. But the U.S. government has few good options. Nobody has proposed a solution that represents a workable compromise to these many interest groups, much less satisfying them. These political problems are likely to continue for some time.

Assurance of Service: Viruses, Worms, and Denial of Service Attacks

The tremendous power of the Internet is seriously compromised by the presence of worms, viruses, and denial of service attacks. Files on individual computers can be corrupted or destroyed, access to the Internet or to specific sites can be denied, unauthorized use of one's computers can happen surreptitiously, physical operations of businesses can be halted, and critical infrastructures such as the banking, electric and water supply, telecommunications networks, and government and defense systems can be threatened. In January 2003, for example, the Slammer worm put 13,000 Bank of America automatic teller machines, all of Continental Airlines' online ticketing and check-in kiosks, and the computer networks of the Atlanta Constitution newspaper out of commission for most of a day. Denial of service attacks are common and growing in frequency. A three-week test in 2001 by the University of California—San Diego identified 12,000 denial of service attacks against 5,000 targets. One organization was subject to over 100 attacks in a single week.

The policy history of dealing with computer viruses is one common to many technological fields that experience rapid innovation and broad social reach—an ongoing battle to catch up both legislatively and programmatically. The first computer crimes occurred in the 1950s, soon after the first commercial computers were installed. It was not, however, until the early 1980s that computer viruses became enough of a problem to warrant congressional action. The first worm was invented at Xerox Palo Alto Research Center in 1979; and the term "computer virus" was coined at the University of Southern California in 1983. Once many people had access to personal computers and Internet connections, viruses became more and more commonplace. Through the early 1980s, the courts tried to apply to computer crimes existing laws for traditional crimes such as trespass and property violations, with limited success. High-profile cases, such as a major bust in 1983 by the FBI of young hackers, known as "the 414s," who used Apple II computers and a modem to break into government networks, popularized the need for new legislation.

The first legislation enacted by Congress came in 1984, the Counterfeit Access Device and Computer Fraud Act. The law applied mainly to computer crime against financial and government institutions. Critics complained that the law was too narrow in the range of institutions it protected, ambiguous because of lack of tight definitions of the relevant computer terms, and unclear jurisdictionally.

To remedy the deficiencies, Congress passed the Computer Fraud and Abuse Act of 1986. It set out six actions as illegal:

1. knowing unauthorized access to national security information,

2. knowing unauthorized access to a government computer with intent to defraud,

3. knowing trafficking of passwords or similar information to defraud that involves interstate or foreign commerce or a federal interest computer,

4. intentional unauthorized access to information from a financial information or a consumer reporting agency,

5. intentional unauthorized access that interferes with government operation of government computers, and

6. intentional unauthorized access to a federal interest computer that results in alteration, damage or destruction of information in the computer or prevents authorized use of the computer or information.

Unauthorized access to a government computer was defined as a felony, while trespass on a government computer was deemed a misdemeanor.

In a high-profile case of 1988, a young programmer named Robert Morris, supposedly out of boredom, released a worm that infected a number of Department of Defense (DOD) computers. Morris was convicted under the 1986 law, fined $10,000, and given a sentence of three years' probation. Despite Morris's conviction, the law turned out to have too many shortcomings to be the truly effective. It was limited to "knowing" or "intentional" actions, so would not cover against negligence or reckless behavior. It applied mostly to computers under federal control and gave only limited interstate jurisdiction. Technical terms were not precisely enough defined. There were no provisions for compensation of property damage caused by a computer virus. Nevertheless, this law has stood as the principal act against computer crime for more than a decade. The Computer Abuse Amendments Act, championed by Al Gore and passed in 1994, extended the law to transmission of viruses. Computer hackers who knowingly or recklessly release a virus that affects computers used in interstate commerce are subject to criminal action of up to 20 years in jail and a fine up to $250,000.

Because of the havoc created by Morris's worm, the DOD established a federally funded research and development center at Carnegie Mellon University, known as the Computer Emergency Response Team (CERT). CERT continues today to be one of the leading players in Internet security. It conducts research and training programs, closely monitors incidents as they occur, and disseminates information about Internet security breaches.

The most recent amendment to the 1986 act was passed in 1996 and is known as the National Information Infrastructure Protection Act. The main thrust of the legislation was to close loopholes that hackers had been using in their court defenses when charged under the 1986 act. The Act fixed a jurisdictional problem, so that now the law covered crimes involving all computers with Internet access. In order to meet the interstate commerce test of federal jurisdiction, past law had applied only to cases in which the computers used in a crime were in multiple states. With this law, the very fact of the interstate reach of the Internet satisfied the interstate commerce test, so long as the computer used in the crime has an Internet connection. Intentionally accessing a computer in excess of one's authority was criminalized, as was to "knowingly cause the transmission of a program, code, or command with the intent to cause damage."

Two laws enacted in the second half of the 1990s, while not specifically addressed to computer viruses, criminalized Internet security breaches in the health care and financial industries. The first was the Health Insurance Portability and Accountability Act (HIPAA) of 1996, which included strong provisions about protecting the integrity and confidentiality of information passing through the health system. HIPAA protects against computer security threats to the information in health systems and against unauthorized uses and disclosures of this data. The second was the Gramm-Leach-Bliley (GLB) Act, passed by Congress in

1999. This act mandated provisions for the security and confidentiality of customer records and information in U.S. financial institutions. Although not as detailed and stringent as the HIPAA regulations, the GLB regulations protect against both computer security threats to the information in health systems and unauthorized uses and disclosures of this data.

In an incident known as Solar Sunrise, because of vulnerabilities in the Sun Microsystems Solaris operating system, more than 500 military, government, and private computers were successfully subject to attack from hackers. At first the government thought that the attack had originated in Iraq, only to later find out that it was caused by two teenagers in California. The federal government was very concerned about these vulnerabilities and in response formed the National Infrastructure Protection Center (NIPC), run by the FBI. NIPC monitors and assesses both physical and computer threats to the critical infrastructures of the nation and assists law enforcement agencies in these kinds of cases. It operates an information-sharing network known as InfraGard.

There has been increased concern about computer viruses and denial of service attacks as a tool of terrorists since the 9-11 terrorist acts in 2001. The Patriot Act strengthens the government's position in dealing with hackers, and some of the agency efforts have been enhanced inside the Department of Homeland Security. This story is told in Chapter 5.

Spam

In June 2003, MessageLabs, a leading provider of email security services, announced that spam—the popular name for unsolicited commercial email—now accounts for more than half (55.1 percent) of all corporate email traffic world-wide. This represented a tremendous growth in spam. For example, Brightmail, one of the leading suppliers of anti-spam technology, indicated that spam represented only 8 percent of all email traffic 20 months earlier. A Harris Poll, taken in November and December of 2002, reported that 80 percent of users found spam "very annoying," 74 percent supported making spam illegal, and more than 90 percent were angered by pornographic spam. These sentiments did not vary significantly by age, income level, gender, ethnicity, or political affiliation.

Why are people so angered by spam? Does the harm go beyond annoyance? Internet Service Providers (ISPs) incur significant costs to try to filter spam and track down spammers. One small Hawaiian ISP, Lavanet, spent approximately $200,000 in 2002 to give spam relief to its 12,000 customers. State and local governments are concerned about the deceptive and fraudulent practices being pitched to citizens. The U.S. Federal Trade Commission reported in 2003 that two-thirds of the spam contained deceptive information, such as false sender addresses or pitched fraudulent schemes to get rich quickly. Parents are worried about the prevalence of pornographic advertisements. A survey by Symantec Corporation showed that 47 percent of children online had received ads for pornographic websites. Businesses are concerned by the productivity loss and expense of having to filter for and otherwise delete spam from their employee's computers. American companies will spend $120 million this year on anti-spam systems and a total cost factoring in worker productivity of $10 billion this year, according to Ferris Research. "Legitimate" direct marketers are concerned about not being able to get their message out over the Internet effectively, tarnished by the less savory practices of other commercial offers.

Why has spam suddenly become such a problem? Email was created in 1972 for use by a small group of graduate students and computer scientists who already knew one another. There were thus strong norms about using email only for scientific and social purposes. These norms were reinforced by policies of the defense department, the originator of the networking technology. Thus in 1978, when Digital Equipment Corporation used the ARPANET directory to send a notice to all ARPANET users on the West Coast announcing receptions to demonstrate the company's new DEC-20 computer, the network administrators chastised DEC and reminded other users of the appropriate uses of the technology.

The term spam was apparently first used in the late 1980s by the online MUD (multi-user-dungeon) community, where individuals chatted and played in multi-person online environments that they created. To "spam" meant to flood the computer with so much information as to crash it, such as entering all at once a large number of objects into the database that represented the man-made environment that the community had created, or monopolizing a chat session by entering a long text entry that had been created off-line and inserted rather than chatting in real time. This was also a community that had strong norms about appropriate behavior and enforced them through social pressure. The term spam had already been used in the 1940s to mean anything commonplace, but the specific use by the MUD community may have come from a 1970 Monty Python skit, in which a person asks a waitress what the restaurant serves, the waitress spiels off a long list of items, all made from Spam, and a group of Vikings sitting in the corner of the restaurant begin singing a meaningless song about Spam.

The Hormel Foods Corp. of Austin, Minnesota, which has sold more than 6 billion cans of spiced ham since it was introduced in 1937, is generally tolerant of the use of the name as a slang term for unsolicited commercial email. However, the company does protect its trademark, for example in 2003 challenging as trademark dilution an attempt by the Seattle software company Spam Arrest to trademark its name as a software vendor and services provider.

A revealing episode in the history of spam occurred in 1988, when a person on a USENET account developed a charity scam, posting to multiple USENET newsgroups asking them to donate to his college fund since he was running short on cash. This message was sent out from Portal, on which anyone could establish an account for a few dollars. Although the term 'spam' was applied to this incident only retrospectively (the term spam was first used in a Usenet context only five years later), there was extensive discussion on Usenet ruing the fact that by then just about anybody could get access to this online communication mechanism, and thus how hard it was becoming to enforce the well-established norms of behavior. This episode presaged the tremendous changes that came about when hordes of people came online through America Online. AOL reached one million members in 1994.

The first major Usenet spam occurred in January 1994, when the Andrews University system administrator sent out a religious message "Global Alert for All: Jesus is Coming Soon" to every newsgroup. There was significant outcry to this behavior, and the systems administrator was sanctioned for his action. The most vociferous complaint was reserved, however, for two lawyers from Phoenix, who in April 1994 sent out advertisements to every newsgroup of legal services announcing their offer of assistance with an upcoming green card (work permit) lottery. The word 'spam' was widely applied to this episode and was used consistently thereafter. The community was particularly outraged in this case because the lawyers were unrepentant about their actions and subsequently wrote a book entitled *How to Make a Fortune on the Information Highway: Everyone's Guerilla Guide to Marketing on the Internet and Other Online Services.*

As the number of users on the Internet grew rapidly through the rest of the 1990s, the commercial opportunities of email became progressively more enticing. With many thousands of unsophisticated users on AOL and other services, the initial norms and culture were increasingly diluted. But it was not until the turn of the new millennium that spamming became prevalent. There seem to be multiple explanations for this: the potential audience has become enormous; the barriers to entry (an Internet connection and some simple bulk mailing software) are low (around $200); and email address lists are readily available (a million names and addresses on a compact disc sells today for about $5). Additionally, high-speed networking makes spamming faster. Some spam-watchers speculate that the market crash has also led marginalized people to supplement their income by spamming, and makes other people more receptive to the kinds of get-rich schemes that spammers so often advertise.

Given the growth in the number and kind of users of the Internet over the past decade, there is no longer a strong set of normative behaviors that the Internet community attempts to impose on individuals who write email or participate in an online discussion. Thus, the main tools for addressing spam are either legal or technical. (There is also vigilante-ism, which will be described briefly below.)

There have been several attempts to enact federal laws to curb spamming, but so far they have not succeeded. This may be because of strong lobbying by the Direct Marketing Association and other trade organizations, as well as the reluctance of a Republican Congress to regulate trade and the reluctance of the Internet community to be subject to government regulation. This situation is beginning to change as the spam problem becomes more pressing, and there are multiple laws

being entertained in Congress. Until recently the federal government left regulation of spam to the states, consistent with the practice of Congress generally to
leave consumer protection matters to the states. To date, approximately half of the
states have enacted laws about spam.

The three states with the most important spam legislation are California,
Virginia, and Washington—not surprisingly, since these are places where there are
heightened levels of activity in the computing and Internet spheres.

In 1998, California Senator Debra Bowen, a Democrat from Redondo Beach,
introduced legislation to control spam. This law, which was enacted that same
year, required spam delivered to California residents by California Internet
Service Providers to place the letters ADV on the subject line in unsolicited
commercial email, and the phrase ADV:ADLT on the subject line in unsolicited
adult email. The legislation also required these spammers to include a toll-free
telephone number or valid email address that recipients can use to remove themselves from the email advertising list. Spammers who do not comply are subject to
a $1,000 fine, but suit can only be brought in district court by a city attorney,
district attorney, the state attorney general, or Internet Service Providers.

Although some individuals had previously sued in small claims court under the
law, the first suit brought by the state attorney general was not filed until 2002. The
state sued PW Marketing, which had sent millions of unsolicited emailemail
messages hawking a "Guide to the Professional Bulk email Business" as well as a
list of 25 million email addresses for $149. Total damages against PW Marketing
could total over $2 million. The case, which was heard in Santa Clara County Supe-
rior Court, was still pending at the time this chapter was written. In December 2002,
Senator Bowen introduced a new bill that strengthened her earlier bill. Bill SB12
bans email advertising unless there is a prior relationship between the sender and
recipient, or unless the recipient agrees to receive such mail.

The bill is modeled after federal legislation that bans unsolicited commercial
faxes. The 1991 Telephone Consumer Protection Act banned using "any telephone facsimile machine, computer, or other device to send an unsolicited advertisement to a telephone facsimile machine." In the case of *State of Missouri v.
American Blast Fax and Fax.com*, the court ruled that the TCPA was unconstitutional on First Amendment grounds, since the Supreme Court had given limited
First Amendment protection to commercial speech in a ruling in 1980. However,
in March 2003 the Eight Circuit Court of Appeals reversed the decision because
the cost-shifting and interference placed on recipients were reasonable for the
government to restrict. This ruling gave hope to people that federal spam law
would stand the constitutional test.

The state of Washington is like California in that it has decided recently to
enact additional legislation to give more clarity in the process and more protection
to the consumer. Washington has been the leading venue for individuals to make
small-claim cases against spammers, and it was the first state to have its anti-spam
law upheld in the courts against First Amendment claims. In Washington, spam

cases can be filed either in small claims court ($4,000 damage limit) or district court ($50,000 damage limit). The new law, which passed unanimously in the Washington House, clarifies that district courts do have jurisdiction over out-of-state spammers, and it gives individuals a right to sue spammers in small claims court for damages up to $500 for messages sent with misleading subject lines, falsified reply information, or disguised transmission paths. The first case filed by the state attorney general under the new legislation was against Jason Heckel of Salem, Oregon, who had allegedly sent 20,000 e-mails selling a booklet on "How to Profit from the Internet." Heckel was found guilty and ordered to pay a $2,000 fine and $94,000 in legal fees. Heckel's lawyer planned to appeal, arguing that the state law violates the protection of interstate commerce in the U.S. Constitution, but we could find no evidence of any appeal having been filed as of the time of this writing.

In April 2002, Governor Mark Warner of Virginia signed what was then the nation's toughest spam law. It enhances previous Virginia anti-spam law, targeting the most egregious of spammers—those who send out at least 10,000 messages within a 24-hour period or 100,000 messages within a 30-day period, or those who generate at least $1,000 from a specific transmission or at least $50,000 cumulatively from these transmissions. Under these circumstances, spamming is a felony, resulting in fines or prison terms of one to five years. Modeled after anti-racketeering laws, the bill permits seizure of all profits and income from the spam advertising. The law applies to email passing through Virginia-based Internet Service Providers, which is significant since half of all U.S. email is routed through Virginia. It is not clear whether the new law will stand up in court, but Virginia's original law had stood up to challenges brought by Verizon Communications and America Online.

Although most of the action so far has been at the state level, there has been increasing federal attention to spam. Bills have been introduced at the federal level each of the past several years, but so far none have been enacted. In 1998, the Federal Trade Commission began to monitor spam. The FTC does not have the authority to bring charges against spammers per se, but it does have the right under the Federal Trade Commission (FTC) Act to prosecute against deceptive business practices conducted through email, such as offers for pyramid schemes or fraudulent business opportunities. The Commission set up an email address (uce@ftc.gov) to which people can send copies of email that they believe to contain unfair or deceptive practices. So far, the FTC has received more than 16 million e-mails and is currently receiving new ones at the rate of more than 50,000 per day. The FTC has established that many of the unsubscribe links in these spam offerings do not work as advertised but instead link the spam recipient to ever more spam lists.

Beginning in early 2002, the FTC began to crack down on spammers whose messages contained deceptive practices. In May, for example, state and federal law enforcement worked with the FTC to make a ten-day sweep, resulting in criminal and civil actions against 45 spammers. The FTC announced a partnership with 21 U.S. and international agencies to close down open relays—those unsecured email servers that spammers exploit to conceal their identities.

In September 2002, three consumer groups (Consumer Action, the National Consumers League, and the Telecommunications and Research Action Center) petitioned the FTC to take harsher action against all spam, not just that with deceptive practices. The commissioners do not feel that they can adopt this wider mission without direction from Congress. In one peculiar twist of events, these consumer groups had established a website <banthespam.com> at which consumers could lodge their complaints against spam. On the website was a button people could click on if they did not want to receive future e-mails from these consumer groups. Unfortunately, the site was programmed incorrectly, and consumers clicking on this button were informed that they would be receiving future email news updates. Thus the antispam site was spamming! The embarrassed webmaster quickly fixed the problem.

In June 2003, the FTC testified before the House Energy and Commerce Committee about spam. The FTC asked Congress to amend the 1994 Telemarketing Act, which protects consumer privacy, so that it would also apply to spam; give the FTC greater ability to counter cross-border fraud; clarify the Electronic Communications Privacy Act so that the Commission could see complaints received from customers by ISPs and to subpoena ISPs in spam investigations; and rule on whether they could reduce the legal protection of spammers who hijack the email addresses of legitimate ISP customers by considering these spammers as unauthorized users. So far, none of these changes has been put into practice.

With the growing amount of spam, interest in federal legislation is growing. However, many people are skeptical that a legal solution will work. Four problems are cited. The multi-national nature of the Internet means that some other country could become a major supplier of spam; and the military, diplomatic, economic, or technological approaches available for stopping them are limited. The techniques of spammers are improving all the time, and legislation that precludes certain kinds of activities by spammers might be circumvented through technological innovation. As both the American Civil Liberties Union and the libertarian think tank, the Cato Institute, have argued, there are also questions of First Amendment rights since certain kinds of commercial speech are protected under the Constitution. Even if the constitutional issues can be resolved, the scope that spam legislation may take is problematic—should it apply to all unsolicited email or only to that which is pornographic, fraudulent, or not providing legitimate headers or ways of opting out of further advertisements?

Four noteworthy pieces of legislation have been introduced recently. In April 2003, Senators Conrad Burns (R–Montana) and Ron Wyden (D–Oregon) reintroduced their Controlling the Assault of Non-Solicited Pornography and Marketing Act (known as the CAN-SPAM Act). Their legislation, introduced in the previous two congressional sessions without passage, was largely responsible for introducing the congressional discussion of this issue. The bill requires spammers to include return email addresses to which recipients can write to opt out of further mailings from that spammer. If the unsolicited email continues, the spammer can be fined $10 per email, up to $500,000, as well as fines of up to $1.5 million for "willingly and knowingly" violating the law. Spammers can also receive jail terms of up to one year for including misleading header information in spam. The bill also requires commercial email to include the letters "ADV" in the header so as to facilitate filtering, and it prohibits spammers from disguising their identities. The previous year, the bill had been watered down in committee in response to lobbying efforts of the Direct Marketing Association. One major change was to eliminate the opt-in provision, which only allows commercial email to be sent if the recipient has requested it or has a previous business relation with the sender. The opt-in provision was replaced with the much weaker opt-out provision described above. State Attorneys General are concerned that the CAN-SPAM Act will preempt state laws, which in many cases are more stringent. CAN-SPAM is still working its way through Congress.

Senator Charles Schumer (D–NY) has also introduced legislation in 2003, the Stop Pornography and Abusive Marketing Act (known as the SPAM Act). In many ways his bill is similar to the CAN-SPAM Act, but it has three interesting differences. It calls for creation of a no-spam registry, modeled after the do-not-call list legislation enacted in 2003 that prevents most calls from telemarketers. It is the first federal bill that allows individuals to sue spammers. It also prevents spammers from "harvesting" email addresses by using software to mine websites and online discussion groups for email addresses. The Schumer bill has received support from an unlikely source. Schumer is one of the most liberal members of Congress—someone who actively supports gay rights and abortion rights, for example. However, the conservative Christian Coalition, which is seldom on the same side of issues as Schumer, has become an active supporter of his SPAM Act, mainly because of the provisions against pornography.

Of the eight pieces of spam legislation currently before Congress, two other bills deserve mention. Representative Zoe Lofgren (D–CA) has introduced the REDUCE Spam Act. This bill includes a provision that would provide bounties to people who track down spammers who do not label their email as advertisements. This bounty approach was suggested by Lawrence Lessig, the Stanford Law professor who is one of the most influential thinkers about Internet and Society issues. Perhaps the bill that has the greatest chance of success is the Reduction in Distribution of Spam Act of 2003 introduced in May 2003 by Richard Burr (R–NC).

This bill has co-sponsorship from two influential representatives: W. J. Tauzin (R–LA), the chair of the House Energy and Commerce Committee, and James Sensenbrenner (R–WI), chair of the Judiciary Committee. It is somewhat weaker than some of the other spam bills and also weaker than national legislation of telemarketing, perhaps because of the influence of email marketing groups such as the

Direct Marketing Association and the National Retail Federation. The legislation calls for opt-out rather than opt-in provisions, allows companies to send advertisements to any individuals for three years following any prior relationship, and allows subsidiary lines of business to be treated separately (presumably meaning that people would have to opt out of advertising from each separate subsidiary of a company).

At a FTC forum on spam in April 2003, Washington State Attorney General Christine Gregoire stated that 44 states plus the District of Columbia would not support the CAN-SPAM Act or the Reduction in Distribution of Spam Act because this would weaken the legislation already in force in 27 states. The states would presumably also have the same issues with the Tauzin-Sensenbrenner bill, which does not allow states to enact tougher laws to combat spam than the federal government. Supporters of Tauzin-Sensenbrenner argue for the need for regulatory clarity, so that all commercial enterprises know the laws and all enforcement agencies can enforce them uniformly.

The direction of federal legislation is not at all clear. One of the most interesting issues to be played out is the definition of spam: is it all unsolicited email advertising, or only that portion that hawks objectionable products (and who defines what is objectionable?), or only those e-mails that do not follow practices such as misleading headers or failure to contain an opt-out option? The Direct Marketing Association, which originally opposed all spam legislation, today supports legislation that bans spam that involves fraud and pornography, because this spam gives a bad name to the email advertisements they would like their members to be able to send out in an unregulated way. But many consumers would prefer to have a much broader restriction on email advertising.

The main alternative to law for solving the spam problem is technology. There are two basic approaches. One involves filtering the email as it is received by the Internet Service Provider, the organizational server, or the individual user's computer. The other involves changing the way in which the email system works.

Filters and blockers installed by Internet Service Providers, such as the software sold by Brightmail, is generally believed to capture between 70 and 90 percent of all spam. Given the large amount of spam, this still lets plenty of spam through the system to its intended destination. Although these filtering systems are routinely improved, so are the techniques of the spammers to foil these filters. Somewhat more successful are filters that are placed on individual computers, such as the open-source Spam Assassin or the commercially available SpamKiller from McAfee. These filters are reported to capture about 95 percent of spam, with only 0.1 percent of "false positives" (blocking email that is not spam). Even better are some of the new Bayesian filters, which observe which messages the individual reads and which she discards, learning from these decisions which incoming messages to block. The filter rate on these is reported to be 99.8 percent with false-positive rate of 0.05 percent. Recently, ICSA Labs, which conducts industry certifications for such products as firewalls and virus protection software, tested a number of open source and proprietary anti-spam products. They found none to have an adequate level of performance to recommend them. But even if most of the spam is blocked at the recipient's computer, the tremendous number of undelivered spam messages is clogging the Internet.

The other solution involves changing the system in some way to discourage spam from being produced in the first place. The simplest approach, the blacklist, is not so different from a filter. Internet Service Providers and nonprofit anti-spam groups such as CAUCE (Coalition Against Unsolicited Commercial Email) and Spamhaus keep lists of spammers, who are prevented access, say, from passing messages of any kind through an ISP. Unfortunately, the ISPs have trouble keeping up with the circumvention techniques, such as regular name changes, that the spammers employ. Some ISPs provide software than enables customers to establish whitelists. These are pre-authorized address books. If the sender is not on the whitelist, the message is not delivered.

Another solution is to charge for the privilege of sending an email. This could be an actual monetary fee, a fraction of a penny per email. Or it could be a "tax" on the sender's email system. A system being considered by some groups, including Microsoft, is to make the sending computer do a small mathematical calculation that ties up the computer for a moment before the email can be sent. Either of these schemes would make it uneconomical for someone to send massive bulk email if the expectation is for only a very small percentage of messages to result in a sale. Such a system offends those who believe in a "free" Internet, and it would require a major change in the protocols for sending email.

Internet Access for the Disabled

The Internet has been widely touted as a force that is a great equalizer in American society, affording opportunities for better government services, a wider range of shopping and entertainment options from one's own home, and greater sources of information. Unfortunately, access to the Internet has not been equally available to all. People in poor and rural communities, as well as most minorities, have not had as wide access as richer, urban, and white populations to either the computers used to access the Internet or to Internet connections themselves. These Digital Divide issues are discussed Chapter 9. Another group that has had limited access is the disabled. The U.S. Bureaus of Census indicates that approximately one-fifth of the U.S. population is disabled in one way or another, and that the percentage is rising with the aging of the Baby Boomers. There are at least two reasons for this lack of access. The disabled tend to be significantly poorer than the population as a whole, and they cannot afford the Internet connections, computers, and often-quite-expensive special devices that help them to use their computers and the Internet. The other reason is that many web sites are not programmed in a way that makes them accessible to the disabled, for example not having text descriptions of graphics that can be read by the special devices for visually impaired users.

This section looks at the legislation, actions by federal agencies, and case law that governs Internet access for the disabled. There is civil rights legislation that provides individuals and groups who have been denied access to particular web sites because of their disabilities the right to sue for remedy. There is also legislation that sets design principles to be used by government agencies. This history is also sprinkled with examples of government agencies using their bully pulpit to get greater compliance with accessibility standards for the disabled. This happens, for example, through amicus briefs filed in court cases by the U.S. Department of Justice, or in the case of Attorney General of Connecticut encouraging a number of tax preparation companies to revise their web pages to conform with the disabled-friendly web pages of the federal and state departments of revenue.

As Table 3–1 indicates, there are significant differences in the Internet practices of the disabled compared with the U.S. population at large.

The grounding law in the discussion of making the Internet accessible to the disabled is the Americans with Disabilities Act, which was signed into law in 1990. Title I of the act concerns discrimination of job applicants and employees who are disabled in both opportunities to work and in benefits and conditions of employment. This part of ADA has not played a major role in the politics or legal issues over Internet access for the disabled. Similarly, Title IV, which concerns provision of telephone relay services, is not applicable here. However, Titles II and III do have relevance.

Title II concerns provision of public services by state and local governments. It requires these governmental organizations to make reasonable changes in their "policies, practices, and procedures" both to prevent discrimination and provide opportunity for equal access and participation. One of the key aspects of Title II for these purposes is the obligation of these governmental organizations to provide "effective communication" with members of the public who want access to information or services. One means mentioned in the law for achieving effective communication is providing auxiliary aids and services, including technological aids (but not specifically Internet aids).

There have been few test cases of Title II. One of the most important involved the California public higher education system. Because of complaints by visually impaired students, mostly under the "effective communication" provision of Title II, the U.S. Department of Education undertook eight investigations of California public colleges and universities. In each case, the investigation resulted in an out-of-court settlement with an individual community college, the state community college system, a campus in the California State University system, or a private university. These cases concerned access to the campus itself, to curricular materials, computer laboratories, and in the case of San Jose State University to Internet access. The settlements in the California cases have been taken as national standards of accuracy, timeliness, and appropriateness in meeting Title II

Table 3–1   Differences of Disabled Internet Use

Disabled more likely to...	Disabled less likely to...
• be newer users of the Internet • have Internet access only from home • look for medical information online • play a game online • research online for information about a particular person • never want to go online • be worried about online pornography, credit card theft, and fraud • believe that the Internet is confusing and hard to use • believe that Internet access is too expensive	• use the Internet • be employed full or part time • be well educated • buy a product online • look for leisure activity information online • have close friends and relatives who use the Internet • believe it would be hard to give up the Internet • have an annual income over $20,000

(Source: Amanda Lenhart *et al.*, "The Ever-Shifting Internet Population, Pew Internet & American Life Project, April 16, 2003. *www.pewinternet.org*)

requirements—not only for education, but more generally in public organizations. These decisions specifically included Internet access.

The other important tests of Title II have involved public transportation systems. The first case that came to court was *Martin et al. v. MARTA*. In this case a group of disabled individuals sued the public transportation system of Atlanta. The federal District Court ruled in 2002 that MARTA has an obligation to provide equal access to visually impaired persons about route and schedule information, including that provided on MARTA's website. The fact that this information was available by telephone to the visually impaired was not seen as equal or timely as the service provided to the general public. The website was accessible 24 hours a day, for example, while the telephone service was available only during limited hours. The court did not direct specific remedies by MARTA inasmuch as MARTA agreed to the inadequacies of its web site and was already in the process of upgrading the system to make it accessible by text reading computer for the visually impaired.

Title III of the ADA concerns discrimination on the basis of disability to goods, services, and facilities of "public accommodation." Essentially, this section deals with situations in which private organizations are legally required to make accommodations for people with disabilities. Political debate and case law have focused on the meaning of "public accommodation." The law lists the private entities that are considered as "public accommodation" under the law. They include hotels, restaurants, theaters, retail stores, services such as laundromats and travel services, public transportation depots, museums, recreation places such as parks and zoos, private schools, social service establishments such as day care centers and homeless shelters, and gymnasiums. The list is remarkable for the fact that, at least on the face of it, it mentions only physical entities. The question is open as to whether this notion of "public accommodation" includes online elements of those physical entities that were specifically listed (for example, the hotel's website), and whether it applies to organizations that provide commercial or other products or services directed at the public but operated entirely online.

The case law has not fully answered the question about applicability of ADA to products and services provided online. In *Carparts Distribution Center, Inc. v. Automotive Wholesalers Association of New England, Inc.*, the federal District Court ruled that the protection under Title III does apply to services of a health insurer. Important to this ruling was the placement of travel services in the list of public accommodations. The court argued that one "can easily imagine the existence of other service establishments conducting business by mail and phone without providing facilities to their customers to enter in order to utilize their services. It would be irrational to conclude that persons who enter an office to purchase services are protected by ADA, but persons who purchase the same services over the telephone or by mail are not." (Carparts, 1994) However, in *Parker v. Metropolitan Life Insurance, Inc.* (1997) the court argued that Title III does not apply to the terms of an employer's insurance benefits program inasmuch as "public accommodation" applies only to places (in this case, the insurer's offices) and not to the services they provide.

The first court case under ADA against a private online organization was brought in 1999 by the National Federation of the Blind (NFB) against America Online. NFB argued that AOL's web browser was incompatible with screen-reader software used by the blind to access AOL. This is software that would render text on the screen into synthesized speech or Braille. The suit also spelled out other features that were available to other users but inaccessible to blind users, such as commands that could have been accessible through the keyboard but were only accessible through the mouse, thereby depriving blind users of access to shopping, entertainment, and other products and services directed at the public (public accommodations). The case was settled out of court, so no case law precedent was established. AOL admitted no wrong-doing, but it established an accessibility policy, consulted with the disabled community over accessibility issues, and incorporated changes that accommodated the needs of vision-impaired users into its version 6.0 software.

When Congress passed the Workforce Investment Act of 1998, this bill represented the most sweeping reform in access for the disabled to the Internet that has been passed to date. One key provision of the act was to implement Section 508 as an amendment to the Rehabilitation Act of 1973. Section 508 spells out minimum standards of accessibility for the disabled to be used in the federal government for software, web-based intranet and Internet information and systems, video and multimedia products, and desktop and portable computers. Access for the disabled to the Internet or information on it has to be comparable to access available for others. The products supplied to the federal government by private vendors must meet these standards; however, the companies themselves do not need to make the equipment used by their employees meet these standards, nor do they need to make their company websites accessible according to these standards simply because they are vendors to the government. Systems designed for military command, weaponry, intelligence, and cryptologic activities are excepted from the standards. Although the law was originally intended only for federal agencies, it has been interpreted more broadly to apply to all state agencies that receive federal funds under the Assistive Technology Act. Thus, for example, state colleges and universities that receive federal funds are required to follow Section 508 standards. Companies providing information technology to the federal government were originally supposed to implement the Section 508 guidelines by March 2000, but President Clinton delayed the implementation to June 2001 at the request of industry so that companies would have more time to study and implement the standards.

The standards in Section 508 for access to Internet service and information were based largely on the Web Accessibility Initiative of the World Wide Web Consortium (W3C). W3C, which is a non-profit coalition formed in 1994 by Tim Berners-Lee, the inventor of the World Wide Web, was one of the first organizations to take an interest in accessibility standards for the disabled (its Web Accessibility Initiative). Its technical guidelines do not prohibit the use of animation or graphics, but adds frames and other metatags that enable text readers to "read" these graphics and translate them into audiotext or Braille.

Throughout the 1990s, the U.S. Department of Justice (DOJ) advocated for better online accessibility for the disabled. One of the most important examples of this was the amicus brief the DOJ filed in the case of *Hooks v. OKBridge, Inc.* (2000) In this case an online bridge club revoked Hooks's online membership for inappropriate postings on its bulletin boards, taking away his right to participate in its tournaments and access to its bulletin boards. Hooks sued under Title III of the ADA, claiming that mental illness was the cause of his behavior. The federal District Court ruled in favor of the bridge club, arguing that the bridge club was a private member organization and thus not subject to civil rights laws, and that the web site was not a place of "public accommodation." The appeals court confirmed the ruling of the district court, but based its finding on the fact that the bridge club could not have known about Hooks's disability and therefore could not have discriminated on this basis.

The DOJ filed an amicus brief in the Hooks appeal, arguing in favor of the Internet as being covered under Title III. The DOJ put forward two main arguments. The first was by analogy to legal decision based on the First and Fourth Amendments. Movies did not exist at the time of the writing of the First Amendment, but freedom of speech rights under the First Amendment surely apply to them. The Fourth Amendment was written before the invention of the telephone and so did not mention the telephone but only "persons, papers, and effects," yet the privacy of telephone conversations are protected under the Fourth Amendment. So why should the ADA not apply to the Internet, even though the Internet did not exist at that time.

The DOJ's second argument was that Title III should be interpreted to mean service *of* a place of public accommodation, not services *at* a place of public accommodation. The DOJ found it "absurd" that a company would be required to offer non-discriminatory services on-site but be allowed to offer those same services in a discriminatory way through their non-physical media of service, such as over the telephone or the Internet.

Two recent court cases conclude the case law pertaining to an accessible Internet. The first one is *Rendon et al. v. Valleycrest Productions Ltd.* (2002) In this case people with hearing and mobility disabilities sued, under ADA, the producers of the television quiz show "Who Wants to Be a Millionaire." The process for being selected as a contestant on the show involved giving rapid answers typed on to a telephone keypad. There was no facility such as TDD that enabled hearing-impaired people to compete to be contestants, and the need for speedy answers made the selection trial unsuitable for people with mobility impairments. The federal District Court ruled in favor of the television producers, finding that the discrimination had not occurred at a place of public accommodation, so that ADA Title III did not apply. The appeals court reversed the decision, arguing that when a procedure clearly precluded access of the disabled, it should be considered as a denial of access and should override the consideration of the place in which the denial of access occurred (over the telephone), so long as the access that was denied was in a place of public accommodation (the television show in the television studio).

In October 2002, four months after *Rendon* was decided by the appeals court, the federal District Court in Florida ruled in *Access Now Inc. v. Southwest Airline Co.* The advocacy group Access Now, on behalf of visually impaired individuals, sued that Southwest's web pages are not accessible to the visually impaired because of the absence of alternative text for graphic information and some other technical features of the website; thus visually impaired patrons do not have equal access to check fares and schedules, receive online discounts, book rental cars and hotels, and learn of special promotions; and therefore the website is a violation of Title III of ADA. The court ruled in favor of Southwest, finding that the website was not a public accommodation. As the judge stated, "Here, to fall within with scope of the ADA as presently drafted, a public accommodation must be a physical, concrete structure. To expand the ADA to cover 'virtual' spaces would be to create new rights without well-defined standards.... Plaintiffs have not established a nexus between Southwest.com and a physical, concrete place of public accommodation." The case was under appeal at the time this chapter was written.

Judge Patricia Seitz in the Southwest case expressed her disappointment in Southwest's efforts to build a more accessible web page, and she expressed her hope that Southwest would make accessibility improvements in the near future. Many companies are in fact sympathetic to making accessibility improvements, both because it is the right thing to do and because it opens up a new customer base. However, the cost can be prohibitive for small- and medium-sized businesses. The Meta Group estimates that it may cost a company up to $200,000 to retrofit an existing website for accessibility, or perhaps one-fourth that cost if accessibility is considered in the original design. Various tools are sold in the marketplace that allow companies to check whether their websites conform with the Section 508 or W3C Web Accessibility Initiative standards. Companies have also been complaining that costs for complying with other federal mandates, such as the HIPAA, the Patriot Act, and the Gramm-Leach-Bliley Act, take money away that could be used for making their web pages more accessible.

One recent development pits spam opponents against disability advocates. A number of major players on the web have been seeking ways to avoid having automated web robots ("bots") visit their sites and either sign up for free email addresses that can be used to send out spam, or collect email addresses from their online databases that can be future recipients of spam. The technique that came into use, primarily in 2003, was for these organizations to impose a verification test that must be passed before one can sign on to the website. These verification tests, sometimes known as Reverse Turing Tests, often involve filling in on a web form a few characters that appear on the web page in a distorted format that is difficult for bots to copy and recognize. Microsoft claims that this practice has reduced its email registrations by 20 percent. Unfortunately, it is also difficult for visually impaired users to recognize these stylized characters. The W3C Web Accessibility Initiative is trying to work out ways that web sites can turn away bots, but in ways that are more friendly to the visually impaired.

Cybersquatting

Companies work hard to build brand recognition among current and potential customers, and protection of their trademarks is one of the most important elements of doing so. Trademarks also protect consumers from unwittingly purchasing shoddy imitations. Over the past decade, with the growth of the World Wide Web, an online presence has become essential to many businesses. As of early 2002, there were 34 million domain names registered and more than two million active commercial websites. In order to be recognized in such a crowded marketplace, companies want to be able to control and be the sole users of domain names that are closely associated with their trademarks. Unfortunately, companies in different businesses but with similar names often vie for the same domain name. Who, for example, should own united.com, United Airlines or United Van Lines? There are cybersquatters who appropriate domain names similar to well-known trademarks in bad faith. A cybersquatter might appropriate a name such as Cadillac.com in order to make a profit by reselling the site at a big profit to the automaker, or take advantage of the online customer who comes to the site thinking he will find luxury automobiles and sell him other products (often porno-graphic). Others might appropriate the domain name in order to prevent Cadillac from having this sales advantage, hence making sales and marketing easier for Cadillac's competitors, or by using the site to lodge complaints about Cadillac products or business practices.

Over the first half of the 1980s the modern system of domain names was devel-oped and placed in operation. Once this system was implemented, users no longer had to know the exact path to websites. Domain names provided an easy, mnemonic way to identify and remember specific locations on the Internet rather than use the actual Internet Protocol (IP) addresses, which are long strings of numbers. The system worked well so long as most of the users were educational and government organizations, and their employees. Problems rapidly appeared once the Internet was commercialized.

Most of the top-level domains, such as .com and .org, are unregulated. What this means is that most anyone can register for any domain name with these suffixes, so long as the domain name has not already been assigned, and the person or organization pays a modest fee, completes a simple registration form, and returns it to one of these authorizing registering organizations. There is no limit on the number of domain names that can be held by any individual or organi-zation. In particular, in order to be awarded a particular domain name, an organi-zation or individual does not need to show that it is entitled to use the name. For example, if some enterprising individual wanted to register Cadillac.com, she would have been able to do so provided that nobody had yet registered that name, without having to demonstrate, for example, that her given or surname was Cadillac or that she was connected with some company that provided products or services that used the Cadillac name. The lack of a test of entitlement for procuring a specific domain name was presumably based on the cost and time of undertaking that test, and the desire of both the Internet user community and the government to make the Internet as widely available as possible. However, as a result of not having any kind of entitlement test, a digital land rush ensued, with many domain names being registered for their potential future value. Many companies that were late to realize the marketing potential of the Internet were locked out from having a domain name that closely resembled their most valuable trademarks; and many disputes over domain names arose.

The principal federal legislation governing trademarks, the Lanham Trademark Act, was enacted in 1946. The Lanham Act prohibited unauthorized use of a trademark on goods or services when such use would be likely to confuse a typical customer as to the origins of the goods or services. Under some circumstances, similar as well as identical marks were prohibited under the Lanham Act. Some domain name registrations would be clearly illegal under the Lanham Act—in those cases where it was clear that trademark infringement or trademark dilution resulted. Trademark dilution is a more recent legal concept, dating from the mid-1990s in the United States. It applies only to "famous" trademarks, and occurs by either blurring the uniqueness of the mark or tarnishing the reputation of the mark.

Unfortunately, the Lanham Act did not provide adequate legal regulation of these domain name issues as they impacted on trademarks. It did not apply to cyberspeculation (registering domain names for resale at a profit at a later time) or to cybergriping (for example, if Cadillac.com had been registered by a cyber-griper, the website would be used to post complaints about Cadillac products, services, or business practices). The act also did not give full coverage against trademark dilution.

Driven largely by concerns about trademark infringement on the Internet, and bowing to international pressure to bring U.S. intellectual property standards more in line with world standards, in 1995 Congress passed the Federal Trademark Dilution Act (FTDA). FTDA closed some of the loopholes in the Lanham Act. For example, it made cyberspeculation illegal under many circumstances; removed one requirement for proving dilution (that the domain name holder be in competition in the same line of business as the trademark holder); and introduced a new kind of dilution, in addition to blurring and tarnishing—dilution by elimination, which is eliminating the trademark holder's ability to distinguish its products and services on the Internet by not having access to the obvious domain names.

Some of the early legal cases based on FTDA involved ideological squatters and arguments about how First Amendment rights to free speech conflicted with FTDA. The most famous of these cases was *Planned Parenthood Federation of America v. Bucci* in 1997. The defendant registered the domain name plannedpar-enthood.com and used it to advertise a book written by an anti-abortion activist, espousing views that were antithetical to those held by Planned Parenthood. The defendant argued that the First Amendment protected use of someone else's mark when it was a case of free speech rather than selling a product, but the court disallowed this argument in noting that the words "planned parenthood" are not a necessary part of the message. The court ruled in favor of Planned Parenthood both on the Lanham likelihood-of-confusion standard and on dilution, and Bucci was enjoined from using the plannedparenthood.com domain name.

The other important legal case of this era was *Bally Total Fitness Holding Corp. v. Faber*. In this case the defendant was sued under Lanham and FTDA for use of the web page compupix.com/ballysucks to express his dissatisfaction with the operation of a particular Bally Total Fitness health club, and for linking this page to a commercial "Images of Men" website that included pornographic images and photographs of gay men. The court ruled against Bally's claim, finding that no reasonable consumer could be confused that the Ballysucks website was associated with Bally Total Fitness. In passing, the court said that this would have been true even if Faber had registered and provided his content on ballysucks.com (not just on a webpage of the compupix.com domain name). Faber had mutated the Bally trademark by printing the word "SUCKS" in large letters across the mark. The court ruled here that the addition of the word "SUCKS" created enough of a difference between the two marks that there was not a reasonable chance of confusion between them. (Other case law, however, has found that alteration of a mark can constitute dilution. In the 1994 case of *Deere & Co. v. MTD Products*, the court ruled that MTD's animated web site on which an MTD product chased John Deere's leaping deer logo around the web page did tarnish the John Deere trademark.) The court concluded that Faber was exercising his First Amendment rights to make public testimony of his critical remarks about Bally's.

Although FTDA helped somewhat, with the growth of the Internet and the dot-com boom in the late 1990s, the shortcomings of the FTDA as a complete solution to the trademark-domain names became ever more apparent. Cybersquatting was a frequent subject of journalistic reports in 1998 and 1999. As a result of all this attention, two solutions were developed in parallel. Congress passed the Anti-Cybersquatting Consumer Protection Act and ICANN implemented the Uniform Domain Name Dispute Resolution Policy within months of one another in 1999. While these two solutions had similar goals, there were very substantial differences in their processes and the penalties they assessed to cybersquatters.

The Anti-Cybersquatting Consumer Protection Act (ACPA) is an extension of the Lanham Act. It applies to people who have a "bad faith" intent to profit from a domain name by "registering, using, or trafficking" in a domain name that is "identical or confusingly similar" to a famous or distinctive trademark. Congress identified in ACPA nine factors to consider in determining whether a domain name registrant had bad faith intent to profit on the registration. These include such factors as attempting to sell the domain name to the trademark holder or someone else without having used it, providing false or misleading information during the registration process, or registering a multitude of names that are identical or similar to trademarks.

Action can be taken under this legislation by any holder of a trademark that is protected under federal law, whether or not the trademark is registered, so long as the trademark is distinctive or famous at the time the domain name is registered. "Distinctive" means that customers have come to recognize it as a particular source of products or services in a particular industry. "Trafficking" means any form of transaction involving the domain name that generates value for the domain name registrant. If the cybersquatter loses a decision under ACPA, the trademark owner can recover the domain name plus either actual damages (the cybersquatter's profits from use of the mark, plus losses to the trademark holder from harm to the mark, plus court costs) or, more typically, statutory damages of $1,000 to $100,000 per domain name.

In late 1999, at almost the same time that Congress passed ACPA, ICANN established the Uniform Domain Name Resolution Policy (UDRP). It was a gesture of cyberlibertarianism; of wanting the Internet community to govern itself rather than be subject to the laws of various countries. The UDRP offered a quick and inexpensive way in which to resolve straightforward domain name issues such as obvious cases of cybersquatting. ICANN holds the authority to assign domain names, and in signing up private organizations as domain name registrars, it required them to abide by the findings of UDRP resolution decisions. Similarly, any person or organization that has registered a domain name with these registrants is automatically subject to the dispute resolution policy.

There are many similarities in what needs to be proved under UDRP as compared to ACPA. For a cybersquatter to lose a domain name through the resolution policy, the trademark holder must prove that she holds the right to a trademark that is identical to or similarly confusing to the domain name, that the domain name holder has no legitimate interest (such as you were legitimately trading in goods under that name) in the domain name, and that the domain name was registered and used in bad faith. ACPA involves taking a case to a federal court; the UDRP works in a different way. ICANN established four organizations (the World Intellectual Property Organization, eResolution Consortium, the National Arbitration Forum, and the CPR Institute for Dispute Resolution) as the resolution providers, and the trademark holder can choose whichever of these four providers she wishes. The provider, sometimes with involvement of both the trademark holder and the alleged cybersquatter, selects a panel of one to three panelists to judge the dispute. Panelists are often lawyers, law school professors, and retired judges, but they need not be. Whereas ACPA cases can lead to fines as well as transfer of the domain name, a UDRP decision in favor of the trademark holder only involves transfer of ownership of domain name and does not impose any fines or damages.

It is instructive to compare ACPA and UDRP. It is much less expensive to file under UDRP than ACPA. It costs between $750 and $1500 to have a single domain case resolved under UDRP, whereas the costs of litigation under ACPA are much higher. A typical case is resolved under UDRP in about forty days, whereas bringing suit in a federal court under ACPA takes much longer. Especially since a trademark owner can bring suit under ACPA if she is not satisfied with the results under UDRP, the vast majority of cases (about 98 percent) are heard under UDRP.

Legal scholars have been critical of the shortcomings of UDRP however. There is less consistency in the UDRP decisions for several reasons: there are no provisions for in-person testimony, filing supplemental documents, or appealing decisions, which are ways that the regular court system hones in on precedents, consistent rulings, and predictable results; UDRP panelists often ignore previous UDRP decisions or existing case law when deciding a case, or they latch on to some particular previous UDRP case as a precedent without necessarily having a good reason for doing so; and some panelists bend the UDRP process in order to reach what they believe is the right decision. Some UDRP providers and some of their panelists are more favorably inclined to trademark rights than others; and there is significant evidence of jurisdiction shopping (i.e., of trademark holders selecting these providers and panelists to decide their cases). In fact, 70 percent of all cases are handled by World Intellectual Property Organization (WIPO), which has a reputation for being pro-trademark. The UDRP penalties are mild (the loss of a domain name registration, which costs less than $100 to register), while the potential financial benefits of cybersquatting are great (trademark holders often are willing to buy the domain name from the cybersquatter for $1000, not to mention income that can be earned by using or leasing the domain name). Thus the UDRP system provides incentive to cybersquat. Moreover, the UDRP process does not protect innocent domain name holders from so-called "reverse domain name hijacking" (see below), for example by establishing procedures that bar people from further use of the UDRP if they have previously abused it.

ACPA and UDRP have proved to be effective tools against the most simple kind of cybersquatting, the cyberspeculation in which the domain name is held simply to be bought back from the trademark holder for a profit. However, ACPA and UDRP have been less successful at controlling four other kinds of cybersquatting: reverse domain name hijacking, cybergriping, typosquatting, and pornosquatting.

Reverse domain name hijacking is the overly aggressive appropriation of domain names by trademark holders, often using legal threat or the UDRP system to obtain domain names that are convenient to have. The most famous story of reverse domain name hijacking concerns <Pokey.org>. This story had a happy ending, but this is more the exception than the rule. In 1997 the lawyers representing Prema Toy Company, which own the GUMBY and POKEY trademarks, sent a letter to Christopher Van Allen to give up ownership of the <Pokey.org> domain name. Van Allen, who had been known as "Pokey" since he was a young child, was at the time twelve years old and had received the domain name as a Christmas present. Van Allen's attorneys replied to the Prema lawyers that the use of the domain name was non-commercial and did not have any impact on Prema's trademark rights. A reply from Prema's lawyers claimed trademark dilution, and a second letter from Van Allen's lawyers observed that a non-commercial site could not lead to dilution. Prema's lawyers then asked Network Solutions to suspend the pokey.org domain name. Internet users from around the world rallied behind twelve-year-old Pokey. Finally, the creator of the GUMBY and POKEY characters relented in "the spirit of Gumby" and had the Prema lawyers retract the request to Networks Solutions.

It is interesting to see how ACPA and UDRP have changed the landscape for cybercomplaint sites. Cybergriping or cybercomplaint sites (of the form <CompanyNameSucks.com>) would most likely be objectionable on the grounds that they diluted the trademark of <CompanyName>. FTDA and ACPA apply to dilution, but UDRP does not specifically address dilution. In fact, the results concerning cybercomplaint sites under UDRP have been inconsistent. In some cases, such as <guiness-beer-sucks.com>, the domain name was awarded to the trademark owner. In other cases, the domain name was allowed to stand either because there was not regarded to be confusion of similarity between the trademark and the domain name, or because of freedom of speech arguments. Through case law, there seems to be a pattern emerging that complaint sites using the name of the trademark <CompanyName.com> are not permitted, but complaint sites with the "sucks" suffix <CompanyNameSucks.com> are permitted. For example, in *Lucent Technologies v. Lucentsucks.com*, the judge ruled that there was not likelihood of confusion of the domain name with the trademark holder. In the case of *PETA v. Doughney*, the judge ruled that Michael Doughney must surrender the peta.org domain name, which he used for a website entitled People Eating Tasty Animals that included materials that were antithetical to the objectives of the better-known PETA organization (People for the Ethical Treatment of Animals).

As one might expect in the innovative world of the Internet, new varieties of cybersquatting are being tried on a regular basis. One example is typosquatting, in which a domain name that is identical to a trademark except for a common typing error is registered in bad faith. The most well-known typosquatter is John Zuccarini, who does not resell his more than 5,000 sites but instead uses them as a place to sell pornography and digital music to the steady flow of people who mistype a web name. In most cases, Zuccarini's domain names meet the test of being confusingly similar to a trademark. Thus it is not surprising that he has lost most of the 56 UDRP complaints and seven cases brought under ACPA. For example, he lost UDRP cases to Microsoft, American Airlines, and Encyclopedia Britannica. But his web sites are very profitable, estimated to bring almost a million dollars a year. He recently lost a case brought against him by the Federal Trade Commission for registering misspelled variations of Victoria's Secret, the Backstreet Boys, and the Wall Street Journal; and he was ordered by the courts to pay $1.8 million in fines. It was thought that Zuccarini had moved to the Bahamas, where he could escape the jurisdiction of the U.S. courts and continue his business, but he was located and arrested by the FBI in September 2003 in Hollywood, Florida, where he had been quietly living for months.

Many Internet users find websites not by typing in the URL but by using a web search engine such as Google to find the site. To take advantage of this practice, cybersquatters have begun to place trademarks in metadata, since many of the search engines look at these metatags in deciding which sites to put at the top of the list of sites found in the web search. Some of these trademarks are placed in metatags in the header code that is attached to the web pages but not shown visually. In other instances, the trademark does appear on the web page, repeated many times, but is not apparent to the person visiting the site because the trademarks are covered over by a field of dark color. As yet, this variety of cybersquatting has not been tested in court.

Another recent trend has been pornosquatting in which purveyors of pornography watch for lapsed domain name registrations (when domain name holders neglect to pay the renewal fee to continue to use the domain name) and sign up to use these domain names, which are often already linked to other sites which send web users to the site. For example, on the "Exploring Utah" website of Senator Orrin Hatch (R–UT), there was a link to a Utah-related search engine to find examples of Utah's scenic beauty. The person who had held the registration for that site had let the registration lapse accidentally, and it had been taken up by a porn-site operator from Connecticut. Thus, the natural beauties that one found on clicking on this search engine were not the ones that Senator Hatch had presumably envisioned.

Cybersquatting in its many forms continue to happen every day. However, the number of complaints brought to UDRP has dropped by almost half in the past two years. The reason is unclear. It may be that the UDRP and ACPA have provided tools to control cybersquatting, or it may be simply an artefact of the dotcom crash and a new belief that having a website is not an absolute guarantee of financial success. UDRP has won the upper hand as an enforcement tool, despite its limitations. In a recent case, the U.S. Court of Appeals ruled in *Dluhos v. Strasberg* that a UDRP decision does not qualify as arbitration under the Federal Arbitration Act. The implications of this ruling still need to be worked out. What is clear is that new technologies and new business strategies will continue to test the limits of the law, and that it is likely that new regulations will need to be put in place eventually.

Further Reading

Orion Armon, "Is This As Good as It Gets? An Appraisal of ICANN's Uniform Domain Name Dispute Resolution Policy (UDRP) Three Years After Implementation," *The Review of Litigation* 22:1 (Winter 2003) pp. 99–141.

Mark R. Colombell, "The Legislative Response to the Evolution of Computer Viruses," *Richmond Journal of Law & Technology* 8 (Spring 2002).

Eric Freeman, "Prosecution of Computer Virus Authors," *Legally Speaking* March/April 2003, pp. 5–9.

Lex Frieden, "When the Americans with Disabilities Act Goes Online: Application of the ADA to the Internet and the Worldwide Web," position paper, National Council on Disabilities, Washington, D.C., July 10, 2003.

Peter Guerrero, "Internet Management: Limited Progress on Privatization Project Makes Outcome Uncertain," U.S. General Accounting Office Report GAO-02-805T (June 12, 2002).

Hans Klein, "ICANN and Internet Governance: Leveraging Technical Coordination to Realize Global Public Policy," *The Information Society* 18 (2002), pp. 193–207.

Amanda Lenhart, John Horrigan, Lee Rainie *et al.*, *The Ever-Shifting Internet Population*, Pew Internet & American Life Project, Washington, D.C.. April 16, 2003.

Milton Mueller, *Ruling the Root*. Cambridge, MA: MIT Press, 2002.

Evan I. Schwartz, "Spam Wars," *Technology Review* July/August 2003, pp. 32–39.

Marcia Smith, "'Junk E-Mail': An Overview of Issues and Legislation Concerning Unsolicited Commercial Electronic Mail ('Spam')" Congressional Report Service, Library of Congress, updated July 11, 2003.

***

Thanks to Indiana University graduate student Jason Gretencord
for his extensive and excellent research assistance on this chapter.

Chapter 4

Internet Use

William Aspray

"The Pentagon's original aim to create an information system that could survive nuclear attack was successful. But not even the computer geniuses of the Pentagon could devise a system to survive the onslaught of lawyers and judges."

—Denis Dutton, "Internet Publishers Caught in Legal Web," New Zealand Herald, July 1, 2003.

T he previous chapter explored the political issues associated with governing the Internet and making it widely available to American citizens. This chapter focuses on political issues associated with the use of the Internet. The first two sections concern two issues of great importance to businesses that want to operate on the Internet: what set of local laws govern their operations, given that the Internet reaches across state borders; and what are the rules for taxing products and services offered over the Internet. The next section concerns the use of computer and Internet voting systems by state and local governments to carry out elections more fairly and with wider voter participation. The final two sections examine two of the most profitable Internet businesses, pornography and gambling, and consider political issues about vice and community standards as they apply to the Internet.

Corporate Uses: Jurisdiction

No business can take the risk of operating without knowing the rules of the game—of what constitutes the boundary between legal and illegal behavior. In different countries of the world, and even in different U.S. states, the local rules differ. The Internet is increasingly the vehicle for what is perceived to be fraud, libel, interstate commerce of pornography, illegal gambling, cybersquatting, and other crimes. Because of its inherently transnational nature, it is unclear which community standards or which set of local rules should be applied. In the case of perceived libel, for example, should the case be tried in the local jurisdiction of the person allegedly libeled, the libeler, the Internet Service Provider of the libeler, the various nodes on the Internet through which the libelous message passed, or some other place altogether? The U.S. legal system has been struggling since the early 1990s to provide an answer to this question through case law. Several different principles have been tried and abandoned, and recent court cases indicate that a consistent approach has not yet been achieved.

The legal foundation for the concept of personal jurisdiction was established by the courts in 1945 in the case of *International Shoe Co. v. Washington*. The guideline set in this case both limits the extent to which organizations and individuals are liable to litigation originating in other states, and restricts the reach of states beyond their jurisdictional limits. The principal is that a court has personal jurisdiction over a nonresident provided that the nonresident has certain "minimal contacts" with the state and that according to "traditional notions of fair play and substantial justice" the nonresident could reasonably "anticipate being haled into court there." If the nonresident purposefully conducted business in the state, for example, this would typically meet the minimal contact standard.

In 1996 a Connecticut company (Inset Systems, Inc.) sought to bring a trademark infringement case against a Massachusetts company (Instructional Set, Inc.) for use of the domain name Inset.com. Instructional Set was using this domain name to advertise its products on the web, even though Inset held the trademark for this name. What was the proper jurisdiction in which to try this case? Although Instructional Set did not maintain offices, employees, or a sales force in Connecticut, the court ruled that the appropriate jurisdiction was Connecticut because of Instructional Set's mere presence on the Internet, which was regarded by the court as a kind of advertising to the citizens of Connecticut. The Inset case has been cited as a precedent in determining jurisdiction in several cases in which the Internet was involved.

A case in 1997, *Zippo Manufacturing Co. v. Zippo Dot Com, Inc.*, set the legal system off in a different direction in determining jurisdiction. This was a case of trademark infringement and dilution, in which the Pennsylvania-based maker of cigarette lighters wanted to stop the California-based Internet news service from using the domain name zippo.com. In deciding jurisdiction in this case, the federal District Court in Pennsylvania argued that simple use of the Internet was not sufficient to establish the minimal conduct standard for jurisdiction. Instead, the court examined the nature and quality of the business conducted over the Internet:

> If the defendant enters into contracts with residents of a foreign jurisdiction that involve the knowing and repeated transmission of computer files over the Internet, personal jurisdiction is proper.... A passive Web site that does little more than make information available to those who are interested in it is not grounds for the exercise of personal jurisdiction. The middle ground is occupied by interactive Web sites where a user can exchange information with the host computer. In these cases, the exercise of jurisdiction is determined by examining the level of interactivity and commercial nature of the exchange of information that occurs on the Web site. (*Zippo Manufacturing Co. v. Zippo Dot Com, Inc.*)

The passive versus active test was applied by the U.S. courts and some foreign courts in jurisdictional decisions a number of times in the late 1990s. However, it was largely abandoned two years later. It was difficult to determine how to apply this test in a consistent and fair manner in many cases, especially since many of the cases involved Internet activity that fell into the middle ground. Some courts questioned the connection between an interactive web site and the presence of actual commerce within the jurisdiction in question. The Zippo test also had the unfortunate policy effect of chilling the development of e-commerce by discouraging the use of interactive web sites.

In 1984 the U.S. Supreme Court had set the precedent of the effects doctrine in *Calder v. Jones*. In this case, a California actress had sued a reporter and editor of the *National Enquirer*, located in Florida, for libel in a California district court. The defendants claimed that they did not have the minimum contacts with the state of California to place jurisdiction there. The Supreme Court ruled, however, that although the article was written and edited in Florida, it was directed at California because that is where the actress lived and worked, and where she suffered the negative effects of the publication.

The effects doctrine was applied in an Internet defamation case, *Blakey v. Continental Air Lines, Inc.*, in 2000. A New Jersey-based employee of the airline sued in New Jersey court for defamation against several of her co-workers, who were writing allegedly defamatory messages from outside of New Jersey to Continental's electronic bulletin board, which was operated by the company from its New Jersey offices. The lower courts ruled against the plaintiff because of lack of personal jurisdiction since the messages were being written from outside New Jersey. The New Jersey Supreme Court reversed the lower court ruling, arguing that although the allegedly defamatory actions were taken from outside New Jersey, the effects were felt within New Jersey and so New Jersey would have jurisdiction. This effects doctrine was applied soon thereafter in several other cases involving trademark or copyright infringement.

One of the shortcomings of the effects test is that it makes an Internet publisher subject to jurisdiction in every state because of the global reach of the Internet. As will be discussed later in this chapter, adult pornography is not illegal, but child pornography and obscenity are. What constitutes obscenity is determined by local standards, so a publisher of adult pornography may be subject to trial in any jurisdiction. What may simply be legal adult pornography in the state in which the publisher resides may be (illegal) obscenity in some other state.

Legal scholar Michael Geist has advocated a targeting approach, in which the determination of jurisdiction considers the intention of the Internet business to enter into a particular state and the actions taken to either enter or avoid that state. Several courts have taken a targeting approach in the past several years. For example, in 2001 the Maryland district court in determining jurisdiction in *American Information Corp. v. American Infometrics, Inc.* argued that the mere presence of a national sales program that was not focused on a particular state but was available in that state did not in and of itself constitute jurisdiction in that state. This principle would apply, the court argued, if the company had a general web site listing an email link to the company, even if the web site and email link might be seen by citizens of that state. It would even be true if, through the national sales program, the company received an occasional inquiry from a customer in that particular state.

Why does all this matter? Two court cases decided in December 2002 illustrate the problem for businesses. In the first case, *Dow Jones v. Gutnick*, Rabbi Joseph Gutnick, a wealthy owner of gold mines in Australia had sued Dow Jones for defamation on account of an article that appeared in *Barron's* magazine (owned by Dow Jones), published in the United States but available online. The suit alleged that Gutnick had been involved in money laundering and tax evasion. Dow Jones had argued that the case should be tried in the United States. But upon appeal the Supreme Court of Australia unanimously upheld a lower court ruling that the case should be heard in Victoria, which is Gutnick's home state in Australia. The Australian courts argued upon effects grounds, that Victoria is where the damage to Gutnick's reputation occurred.

This ruling has had company executives and Internet scholars concerned because the conservative libel laws of Australia strongly favor the plaintiff while the United States has one of the most liberal libel laws of any nation, especially as regards public figures. Such a ruling would mean that a publisher would have to pay attention to the libel laws of all 190 countries that have Internet access. Already, the Internet legal community was nervous about this issue because earlier in 2002 the Zimbabwe government had criminally charged an American journalist with "publishing a falsehood." The reporter had written an article in the British newspaper *The Guardian* (only available in Zimbabwe online) alleging ties between President Robert Mugabe and suspected murderers. The reporter was eventually acquitted, but the case was widely seen as harassment of freedom of speech.

In the second court case (*Young v. New Haven Advocate*), decided only three days after the Supreme Court of Australia ruled in the Gutnick case, the U.S. Court of Appeals in determining jurisdiction took a position diametrically opposed to the Gutnick case. Two Connecticut newspapers, the *New Haven Advocate* and the *Hartford Courant* had published articles in print and online about the practice of sending Connecticut inmates—many of whom were black or Hispanic—to the maximum-security Wallens Ridge State Prison in Big Stone Gap, Virginia because of overcrowded Connecticut prisons. The articles noted that Connecticut prisoners had complained about racist remarks and unnecessary use of force from the Virginia prison guards. Stanley Young, the warden at Big Stone Gap prison, sued for defamation. The lower courts had given jurisdiction in Virginia on effects grounds. The appeals court reversed the decision and threw out the defamation case in Virginia on targeting grounds. It argued that the two newspapers targeted their web sites at Connecticut readers with local news stories, weather, and classified ads; and that the Connecticut newspapers did not have sufficient contacts with Virginia to support jurisdiction in Virginia.

The most famous Internet jurisdictional case has been *Yahoo!, Inc. v. La Ligue Contre le Racisme et l'Antisemitisme*. According to the French Penal Code, it is illegal to sell Nazi memorabilia in France. The web company Yahoo! had obeyed French law on its French web site *<www.yahoo.fr>* but it had allowed such sales on its general website *<www.yahoo.com>*, which is directed primarily at the United States market. Indeed, 1200 Nazi items were offered for sale on the U.S. Yahoo site. Three French civil rights organizations, the Union of French Jewish Students (UEJF), the International League Against Racism and Anti-Semitism (LICRA), and the French Movement Against Racism (MRAP) sued in the French court system, arguing that French citizens had full access to the Yahoo.com site. The French judge ruled in May 2000 that Yahoo! must block access by French citizens to all its sites where Nazi memorabilia are sold. In August 2000, the judge charged a British, French, and American expert to look into ways to block the Yahoo.com site from web surfers in France. They reported back in November that it was straightforward to block the site from about 70 percent of French users and that the additional use of a filtering system might allow blockage of up to 90 percent of the population; and two weeks later the French judge ruled that Yahoo! must comply with the measures suggested by the technical advisors within 90 days or pay a daily fine of about $13,000 for each day our of compliance.

In January 2001, Yahoo! banned the sale of Nazi items on its U.S. website. It claimed that its action was based upon consumer request, not on the decision by the French court. At the same time, it went to a California court seeking a declaratory judgment that any damages imposed by the French court would not be enforceable in the United States. The California court ruled in Yahoo!'s favor, arguing that the French court decision violated both the First Amendment right to free speech and the Communications Decency Act's provisions that protected Internet Service providers from liability for third-party content carried on their Internet service. The French civil rights groups appealed the California court decision, and the oral testimony was taken in this appeal in December 2002. No decision had been made at the time of this writing.

Thus, there is great uncertainty in the American legal system over which principles should guide decisions about jurisdiction in Internet cases, and thus about which local laws should apply. Should the passive versus active, effects, targeting, or some other principle be used to decide these cases? There is also uncertainty about enforceability of court rulings upon individuals and organizations in other states.

Corporate Uses: Internet Taxation

As late as 1994 it was against strongly set social norms to use the Internet for commercial purposes. But increasingly over the past decade the Internet has been used to sell at every point of the supply chain, to businesses and individuals alike. Forester Research claimed that Internet sales totaled $79 billion in 2002, about 3 percent of all retail sales in the United States. States and local governments often do not earn tax revenue from Internet sales in the way they do from "brick and mortar" sales. Sales and use taxes are a main source of revenue for 45 of the 50 states, plus the District of Columbia. With the downtown of the economy in the past several years, there is new incentive for the states and local government to find ways of taxing Internet sales; and thus it is not surprising that there have been concerted efforts to do so.

Traditional merchants are concerned about the competitive disadvantage they work under if Internet sales are not taxed but sales in a physical store are. Some policy analysts are concerned about Digital Divide issues: Internet access continues to be more widely available to wealthier and white families than to poor or minority families, so if there is an economic advantage to buying online, it is disproportionately available to some sections of the American public. A substantial number of the online sales have been to gambling and other forms of entertainment, software, and airline tickets, which are often not subject to sales or use taxes. Despite these limitations to large benefits from Internet taxes, states see the Internet sales growth potential and want a piece of the action.

There are three basic political issues: (1) whether to tax the Internet; (2) what products and services to tax; and (3) whether to collect use taxes. Much of the controversy has been over use taxes and the closely related concept of nexus. A use tax is a tax levied by a state on a purchase not made within the state. For example, if a resident of Virginia made a mail-order or Internet purchase from a vendor outside the state and did not pay Virginia sales tax at the time of the purchase, the resident is legally obligated to pay a use tax (at the Virginia sales tax rate) to Virginia. This use tax is typically collected as part of the filing of individual state income tax. The problem is that most individuals do not comply with the state law and pay these use taxes. Moreover, it is very difficult for the state to enforce the payment of use tax by individuals. Thus the state would like the vendor to collect the use tax at the time of purchase.

The Internet merchant prefers not to have to charge a tax to its customers (and thus increase the cost to the customer). It is also expensive to keep track of the tax laws of the 16,000 state and local municipalities, and to pay each of them the correct amount of tax on a timely basis. By law, in-state sellers have to collect sales tax, while out-of-state sellers have to collect use taxes if they have nexus (i.e., substantial presence) in the state. What constitutes nexus is a complicated issue that has been subject to considerable case law. An out-of-state seller with a warehouse or retail store in Virginia would have nexus and would have to collect use taxes. But what if the company merely has sales representatives who visit the state, or the company merely advertises in the state but does not have a physical presence, or the company has a relationship to another company (e.g., Barnes and Noble and BarnesandNoble.com) that has a presence in the state but does not itself have a presence itself?

Two court cases set the context for rulings on nexus and for all subsequent legislation on Internet taxation (as well as taxation on phone and catalog sales). In *National Bellas Hess v. Department of Revenue of Illinois*, the U.S. Supreme Court ruled in 1967 that a certain minimal level of contact must exist for a state to tax an out-of-state business, and that by having a sales office or sales personnel in the state was sufficient to constitute nexus. In 1992, in *Quill Corporation v. North Dakota*, the court ruled that if an out-of-state company purposefully takes advantage of the economic market in a given state in order to sell its products there, it may be subject to the tax-collection provisions of the state even though it has no personnel or physical facilities in that state. However, the court also ruled that the complexity and variety of state and local tax systems created an undue burden on interstate commerce for the out-of-state company. Thus while Quill was not required to collect sales or use taxes for North Dakota, the courts left open the possibility that out-of-state companies would have to collect use taxes in the future if the state and local tax systems were made less complex in a way that lessened the burden on companies to collect and pay these taxes.

The Quill ruling left an opening for Congress to reform the use tax collection system. Senator Dale Bumpers (D–AR) introduced legislation concerning mail-order companies and use tax three times in the mid-1990s, but it was never enacted. The National Governors Association adopted a policy statement, the Internet Development Act, which followed the outlines of Bumpers bills. The Governors' policy called on states to simplify the tax structure by establishing a single state-wide tax structure for mail-order and Internet sales, as well as a simple administrative structure for collecting taxes, in return for the opportunity to force remote sellers to collect use tax. The Governors' policy also called for a moratorium on any new local, state, or federal taxes on Internet access—so as to encourage the growth of electronic commerce.

In 1998, Congress passed the Internet Tax Freedom, Act, sponsored by Rep. Christopher Cox (R–CA) and Senator Ron Wyden (D–OR). This law placed a three-year moratorium on the right of states and local governments to place new taxes on Internet access, such as on the fees paid by customers to their Internet Service Providers. The law also prohibited multiple or discriminatory taxes on electronic commerce (defined in the law as sale, lease, license, offer, or delivery of products, goods, services, and information over the Internet). Thus, under the ban on multiple taxes, two different states could not tax the same e-commerce transaction, but it is permissible for a state and a local municipality to tax the transaction. Under the ban on discriminatory taxes, it is not permitted to tax Internet sales in different ways or at a different rate than sales conducted in person, by mail order, or over the telephone.

The Internet Tax Freedom Act of 1998 had a sunset clause, under which the law expired in October 2001. There was political debate in 2001 over whether to extend the 1998 law, pass a revised version, pass some new law, or take no legislative action.

One issue in the debate concerned extension of the moratorium on new taxes on Internet access, and on how to handle those state and local taxes on Internet access enacted prior to October 1998, which were grandfathered under the original Internet Tax Freedom Act. Several groups favored no extension in the moratorium. They included groups that wanted no federal restrictions on the rights of states and local municipalities to tax, those who felt that such a moratorium was premature since there were relatively few state and local governments planning such taxes, and those who wanted to tie the moratorium on Internet access changes to federal endorsement of state and local rights to require out-of-state companies to collect sales or use taxes. This latter group was concerned about the uneven playing field between Internet companies and bricks-and-mortar companies. Some groups interested in resolving the sales and use taxes issue favored a temporary extension of the moratorium, for no more than two years. This, they believed, would be enough time for the state tax simplification schemes, which were a prerequisite for federal endorsement, to be worked out and implemented. Some groups preferred a permanent moratorium on Internet access taxes. Groups holding this position were either opposed to taxing the Internet because they believed the Internet should not be subject to federal regulation or because they wanted to keep the cost of Internet access low in order to promote the use of the technology and make it more easily affordable to low-income groups. These groups were also in favor of no longer allowing the Internet access taxes to remain in force in ten states that had been grandfathered in the 1998 law.

The states argued that, if the federal moratorium was extended, the definition of Internet access should be narrowed. In the late 1990s, Internet Service Providers had not only provided access to email and the Internet, but had also packaged various services with the basic access, such as information services, cable television access, and online books and music. Thus the state wanted the language of any new legislation to make clear that the moratorium on Internet access taxes did not limit the states' rights to tax this content that was being bundled with the basic service.

None of the proposed legislation in 2001 called for the states to be compensated if the grandfathered access fees were no longer to be permitted. The Congressional Budget Office estimated that the loss of revenue to the states was to exceed the cost threshold ($56 million per year in 2001) that triggers the provisions of the Unfunded Mandates Reform Act of 1995, making it more difficult to pass the bill.

There was one other issue concerning Internet access charges in the 2001 political debates, so-called "Internet kiosks." The 1998 law had been written in a way that prevented nexus from being established merely by the ability in-state to access a website on the out-of-state seller's computer server. Some companies had taken advantage of this language by incorporating their Internet business as a separate legal entity from their brick-and-mortar company and installing Internet kiosks in their physical stores, where customers could make online purchases with their Internet sales affiliate without having to pay taxes. States wanted this practice to be redefined as creating nexus and thus subjecting the Internet sales organization to collection of sales and use tax.

All of the issues in the political debate of 2001 discussed so far in this section have concerned Internet access taxes. The other major topic of debate was whether Congress was willing to encourage states and local municipalities in the tax simplification and implementation strategies. Congress could do this by indicating that it would use its authority over interstate commerce to give the states the right to force out-of-state businesses to collect use taxes from customers at the time of sale. Representatives of high-tech states, which were likely to gain more from the growth of Internet businesses than from extra use tax revenue, tended to oppose federal endorsement of use tax collection. Most other states were in favor (however, in a striking reversal, Governor Gray Davis of California has agreed to consider the Internet tax plan, given the state's precarious financial situation and the diminished influence of the high-tech industries since the dot-com crash).

Several bills were introduced in 2001, representing various combinations of the positions on extension of the Internet access tax and support for the use tax collection. Perhaps not surprisingly, when faced with so many contrary pulls, Congress chose to extend the 1998 moratorium for two more years, until November 2003, without change. The 2001 extension did not address the use tax issues.

With the tax extension of 2001 about to expire in late 2003, there is again political debate of the issues. The issues and positions are much the same as they were in 2001, but there has been some change in the context. Almost all states are facing serious budget deficits, so they are pushing hard for the new revenues that could come from time-of-sale collection of use taxes. To gain federal approval for use tax collection, more than two-thirds of the states have entered into the Streamlined Sales Tax Project.

The states initiated the Streamlined Sales Tax Project in March 2000. The goals are to simplify the administration of state sales and use taxes and to develop or endorse software and a payment system that will be cost-effective for out-of-state businesses to employ. A model act and agreement were drafted in January 2001 for states to take back to their legislatures. At a conference in November 2002, 34 states plus the District of Columbia approved a model interstate agreement. This agreement includes uniform definitions to use in state tax laws, plus a single rate to be used throughout the state for each type of product (to be phased in by 2006). The act becomes effective as soon as at least 10 states satisfy the agreement, provided that those states together representing at least 20 percent of the population of the 45 states that collect a state income tax. As of June 2003, 11 states had ratified the agreement (with certification pending in 12 additional states), but they did not together represent 20 percent of the population. So the Streamlines Sales and Use Tax has not yet come into effect.

Even when the Streamlined Sales and Use Tax is ratified by enough states, as it seems certain to be, use tax collection will likely face serious opposition in Congress. Although Senator Byron Dorgan (D–ND) is planning to introduce legislation that would give federal endorsement to the Streamlines Sales and Use Tax, there are powerful Republicans in Congress who are opposed, and the position of the Bush Administration in not clear. Many small Internet businesses are complaining that collecting use taxes would be onerous to them, and the National Retail Federation, which represents many of the large retailers, is calling for an exemption for small businesses. Others, such as Americans for Tax Reform, are concerned about consumer privacy when all of these online transactions are being tracked.

One recent wrinkle in this unfolding saga is an amnesty program that some large retailers, including Wal-Mart, Marshall Fields, Target, and Toys R Us, entered into with 38 states and the District of Columbia in January 2003. These companies agreed voluntarily to collect sales and use tax from customers in all 45 states that collect sales tax in exchange for amnesty against liability for not collecting use taxes in past years. Companies such as e-Bay, Amazon, and Dell continue to resist collecting use taxes; although the effect of Amazon's sales and marketing partnerships with Target and Marshall Fields on Amazon's obligation to collect use taxes is unclear. Some large retailers such as Sears, Gap, and Circuit City have been collecting sales and use taxes all along. Some states have not been willing to go along with this amnesty program. Illinois threatened in February to sue several large online retailers, including Office Depot, Target, and Wal-Mart, for past use taxes, and New York State is considering following suit.

Government Uses: E-Voting

Over the past decade, as described elsewhere in this chapter, the Internet has been used for legitimate commercial purposes and for providing a new outlet for vices such as gambling and pornography. Although federal, state, and local governments have not been as fast to embrace the Internet as commercial interests have, in recent years government organizations have begun to use the Internet to make their operations more efficient and cost effective as well as to make government services more readily available to citizens and businesses. It is beyond the scope of this book to tell the story of the rise of digital government in America, for that is not primarily a story of information technology policy. However, one small but important part of digital government—e-voting—is closely associated with IT policy. E-voting is not yet primarily about voting on the Internet, although that is one of the emerging policy issues. But for convenience of organization, the entire story of e-voting including Internet voting is told here.

Since the early 1990s there has been a vision that the Internet and computer technology will increase public participation in elections. People with various disabilities will be able to vote on disability-friendly devices from their own homes. Military personal and others away on travel will be able to vote more quickly and easily over the Internet. Other busy people will be able to vote from home or work and not have to take time out to visit a physical polling site. The vision has run well ahead of the technology or of government will to implement any kind of e-voting. In the 1996 presidential elections, for example, the major candidates all had web sites. But they were rudimentary and essentially contained little more than electronic copies of printed pamphlets.

In the early 1990s, the majority of voting was done by filling in paper ballots with pencil and counting them manually, by setting switches on a mechanical lever machine that kept a running tally of votes, or by using a metal punch to punch cards that were sorted and tallied on either specialized punch-card equipment or on a computer. By the year 2000, the situation had changed dramatically. Only a third of voting was conducted by paper ballots or mechanical means. The most common method, used by about 40 percent of all U.S. counties, was to have the voter fill in a circle or complete an arrow, which could then be optically scanned for tallying votes, Direct-recording electronic devices were also beginning to appear. With them, the voter pushed a button or touched a screen and the machine recorded an electronic mark and kept a running tally electronically.

It was not only the method of voting that had changed by the time of the 2000 elections. Use of websites by individual political candidates and by political parties had become widespread. Websites not only were more sophisticated in their form and content, giving daily updates on the campaign, they were also used to help build up databases of friendly voters and solicit campaign contributions. Web-based organizations, such as Pseudo.com, covered the elections in great detail, including online chats with candidates and exclusive web coverage of political events and behind-the-scenes interviews.

The Department of Defense conducted a Voting Over the Internet pilot project in the November 2000 election. 350 voters who were registered to vote in 14 counties from the states of Florida, South Carolina, Texas, and Utah and who were eligible to cast absentee votes under the Uniformed and Overseas Citizens Absentee Voting Act were given the option to sign up and vote online. 91 people signed up and 84 votes were cast over the Internet from twelve countries and 28 states. The review of the pilot noted some advantages for military personnel on assignment away from home: easier access to ballots for on-the-move personnel, and response time much faster than mailing in an absentee ballot. However, the review also noted the cost (approximately $74,000 per voter), serious issues about ensuring the integrity of the voting process when scaled up, and problems that even computer-savvy military personnel had in using digital signatures to authenticate their votes.

Earlier that year, in March 2000, the Democratic Party of Arizona allowed Internet voting in its primary elections. All 843,000 registered Democrats were mailed a personal identification number. With that number and two additional pieces of personal identification, they could vote online at the Democratic Party's website, either from a remote site between March 7 and 10 or at the polling site on March 11. People could also cast a traditional paper ballot at the polling place on March 11. About 41 percent of the votes were made remotely on the Internet.

The New York-based company, Election.com, which ran the election for the Democratic Party, regarded it as a major success; but some problems were reported. Some Macintosh computer users could not vote because of incompatibility of their system software with Election.com's security software. A number of registered Democrats lost or did not receive their PIN numbers, making it impossible for them to vote online. The non-profit organization, the Voting Integrity Project, filed a lawsuit in U.S. District Court against the online voting, arguing that the practice violated the Voting Rights Act because minorities were less likely to have computer access and thus less likely to vote in this convenient manner. The judge refused to block the primary and the Voting Integrity Project has since been disbanded.

However, the hallmark event of the 2000 elections for e-voting was the controversy over tallying of the vote in Florida, upon which hung the outcome of the Presidential election between Al Gore and George Bush. Because of problems with the ballots completed on the old punch-card voting equipment and other election irregularities, it was not until December 12, five weeks after Election Day, when the U.S. Supreme Court ruled 5-4 in favor of George Bush's motion to suspend further recount of disputed presidential election votes in Florida, that the outcome of the presidential election was decided. The following day, Gore conceded and Bush became president elect.

In order to avoid the embarrassment, expense, public distrust, and potential constitutional crisis of another election disputed because of the voting technology employed, both the House and Senate introduced legislation to reform election technology. The general belief was that the problems of the 2000 presidential election could be overcome by replacing older paper and manual technologies with new electronic, computer-based voting technologies. The House passed a bill in December 2001 and the Senate in April 2002. After considerable wrangling over details (mainly over how to enforce federal guidelines), a compromise was reached and the Help America Vote Act was passed in October 2002 and signed into law.

The Help America Vote Act provided $3.9 billion to the states over three years, mostly for new voting machines to replace the mechanical machines that had created the problems in the 2000 presidential voting in Florida. The bill also provided funds for developing computerized databases to give more accurate voter lists. These computerized vote registration databases are supposed to be able to cross-check with other government-held databases such as tax driver's license, and Social Security rolls. The bill also pays for at least one machine in each precinct that will be accessible to people with disabilities. Smaller amounts of money were set aside for developing and testing new voting technology and training election workers.

The effect of the new legislation was to create a major business opportunity for the manufacture and sale of voting machines. Four major studies published in 2001 concluded, however, that Internet voting placed the integrity of elections at too great a risk at this time: the National Workshop on Internet Voting (organized by the Internet Policy Institute and sponsored by the National Science Foundation), the CalTech/MIT Voting Technology Project, the National Commission on Federal Election Reform (chaired by former presidents Gerald Ford and Jimmy Carter), and Georgetown University Public Policy Institute's Constitution Project. In April 2002, the highly influential Federal Election Commission approved new Voting System Standards that also encouraged avoidance of remote Internet voting.

Punch-card systems, with their propensity for hanging chads, were seen as outmoded and prone to error. So the new focus was on voting machines that employ optical scanners or touch screens. New products were offered in the marketplace by Diebold, Election Systems & Software Inc., Sequoia Voting Systems, Hart InterCivic, MicroVote, Avante, Voting Technologies International, VoteHere Inc., and others.

The states were generally unhappy about how long it took Congress to allocate funds to them for new voting machines, and most states had not replaced their old voting machines in time for the mid-term elections in November 2002.

However, four states that were particularly concerned about this problem—Florida, Georgia, Maryland, and Texas—have spent considerable state funds to put in place electronic voting systems by then. These states provided a test of the new voting systems; the results were not encouraging in any of these states.

In Florida, for example, in a March 2002 city council election in Wellington (near Palm Beach), the vote tally was 1,263 to 1,259, but the new Sequoia voting system failed to register 78 votes. Inasmuch as this was the only election on the ballot, it is likely that the 78 people who voted but whose votes were not registered by the voting machine had indeed voted in this particular contest.

On primary day, there were problems booting up machines at the beginning of the voting day, keeping them from crashing during the day, a number of machines that never functioned, and many voters who were sent inaccurate information about where to vote or who received incorrect voter registration cards. In Miami-Dade County, at 9:45 A.M. 68 of 754 precincts had not yet opened because of problems with the new voting equipment. At 10:50 A.M., 32 precincts were still closed and 45 were operating at only half capacity. Some people got tired of waiting and gave up on voting. Under pressure from the Democratic gubernatorial candidate, Janet Reno, Governor Jeb Bush declared a state of emergency and kept the polls open for two extra hours. While most of the problems were with touch screen voting terminals, there were also problems with the optical scanning systems in Union County, which only recorded the votes for the Republican candidate and had to be counted by hand. A few machines were misprogrammed with the wrong ballots, and there were numerous problems and delays in tallying results using the new machines.

Similar problems, although less extensive, occurred in the regular elections in November. For example, a programming error caused the automated system to throw out 34,000 votes and an operator error eliminated another 70,000—both in Broward County—caught the next day because these errors caused a report of a low (34 percent) voter turnout.

When confronted with these problems, the vendors of the voting machines defended themselves, attributing the problems mostly to human error and to a few small programming glitches that were easily fixed. However, after the 2000 elections the computer science community began to speak up more stridently about the problems not only with the electronic voting machines currently on the market, but also with totally electronic or Internet voting in principle. These arguments were based on their experiences with building other kinds of computer systems. They argued that electronics or Internet voting is inherently untrustworthy.

One of the leading critics of fully electronic and Internet voting systems has been Rebecca Mercuri, a professor of computer science at Bryn Mawr College and a consultant to the Gore campaign on the Florida recount vote in the 2000 presidential election. Mercuri has identified ten problems with electronic voting (see *www.notablesoftware.com*):

1. There is no way to verify that the vote cast corresponds to the vote recorded, transmitted, or tabulated.

2. Electronic voting systems that do not have individual print-outs for examination by the voters do not leave an independent audit trail so that the results can be checked.

3. There are no standards from any government or independent standards organization for verifying that any particular voting system is secure.

4. There are no required standards on voting displays, so there is no reason to believe any of the new products will be less confusing than the systems that were used in the Florida 2000 presidential election.

5. Electronic balloting reduces the opportunities for election officials and challengers to perform checks on the veracity of results and concentrates the veracity of the system in the hands of the few people who build, program, and maintain the machines.

6. Although convicted felons and foreign citizens can not vote in US elections, they can and do work for—and even own—voting machine companies supplying equipment for U.S. elections.

7. Encryption systems can often be broken, making it possible to identify the votes of individual people.

8. Internet voting opens up the possibilities of worms, viruses, and distributed denial of service attacks that can either shut down the voting system or compromise its integrity.

9. Internet voting is less available to the poor, those living in rural areas, the elderly, and the disabled.

10. There is not an adequate system for authenticating voters who vote remotely through the Internet, opening up problems of privacy, vote-selling, and coercion.

The vendors aggressively rebutted the criticisms of the computer scientist and steadfastly refused to let the computer scientists examine the software or physical interior of their voting machines, citing trade secrets and security risk. However, a version of the source code for the software that runs the Diebold AccuVote—TS terminal, was posted on the Internet. This voting technology was used across the state of Georgia in the 2002 elections, and Maryland had signed a contract to purchase up to $56 million of these touch screen terminals. Three researchers from the Information Security Institute at Johns Hopkins University, together with a computer science faculty member from Rice University, conducted an evaluation of Diebold's system. Their findings (quoted below) were troubling, despite the detailed rebuttal from Diebold.

Even with this restricted view of the source code, we discovered significant and wide-reaching security vulnerabilities in the Accu-Vote-TS voting terminal. Most notably, voters can easily program their own smartcards to simulate the behavior of valid smartcards used in the election. With such homebrew cards, a voter can cast multiple ballots without leaving any trace. A voter can also perform actions that normally require administrative privileges, including viewing partial results and terminating the election early. Similar undesirable modifications could be made by malevolent poll workers (or even maintenance staff) with access to the voting terminals before the start of an election. Furthermore, the protocols used when the voting terminals communicate with their home base, both to fetch election configuration information and to report final election results, do not use cryptographic techniques to authenticate the remote end of the connection nor do they check the integrity of the data in transit. Given that these voting terminals could communicate over insecure phone lines or even wireless Internet connections, even unsophisticated attackers can perform untraceable "man-in-the-middle" attacks.

As part of our analysis, we considered both the specific ways that the code uses cryptographic techniques and the general software engineering quality of the construction. Neither provides us with any confidence of the system's correctness. Cryptography, when used at all, is used incorrectly. In many places where cryptography would seem obvious and necessary, none is used. More generally, we see no evidence of rigorous software engineering discipline. Comments in the code and the revision change logs indicate the engineers were aware of areas in the system that needed improvement, though these comments only address specific problems with the code and not with the design itself. We also saw no evidence of any change-control process that might restrict a developer's ability to insert arbitrary patches to the code. Absent such processes, a malevolent developer cold easily make changes to the code that would create vulnerabilities to be later exploited on Election Day. We also note that the software is written in C++. When programming in an unsafe language like C++, programmers must exercise tight discipline to prevent their programs from being vulnerable to buffer overflow attacks and other weaknesses. Indeed, buffer overflows caused real problems for AccuVote-TS systems in real elections. (Tadayoshi Kohno, Adam Stubblefield, Aviel D. Rubin, and Dan S. Wallach, "Analysis of an Electronic Voting System", July 23, 2003, *http://avirubin.com/vote/.*)

Although the criticisms of electronic and Internet voting systems were quite serious, the computer scientists had some difficulty in getting their views heard with the people who make decisions about elections. First, the difficulty arose because there was a widespread belief that there was a serious problem with the old punch-card voting systems, which it was imperative to replace; whereas the problems being raised by the computer scientists over the new voting systems was regarded as potential and speculative rather than actual problems. Second, the lack of hearing was a result of the vendors of the new voting systems having well-funded marketing and lobbying efforts bent on assuring politicians that their systems would work well once some small kinks were ironed out. Third, there was difficulty because the concerns were over highly technical issues in code that were hard to show or explain to politicians.

The computer scientists finally did get their hearing in 2003. In March 2002, California passed a referendum, the Voting Modernization Bond Act, that provided $200 million for counties to replace or upgrade their voting equipment. The Help America Vote Act passed by the federal government as well as some other state legislation called for voting equipment that was more friendly to persons with disabilities. In February 2002 a federal judge ordered that all punch-card voting systems in California be replaced by January 2004. Thus counties began to consider purchase of direct recording electronic voting equipment, such as touch-screen systems, to replace paper ballots and mechanical punch-card systems.

One of the counties that planned to replace its voting equipment was Santa Clara County, home of Stanford University and much of Silicon Valley—a place where the politicians as well as the voters were tech-savvy. At the recommendation of county staff, Santa Clara County was just about to enter into a $20 million contract to 5,000 touch-screen voting machines from Sequoia Voting Systems of Oakland, California, until Stanford computer scientist David Dill appeared before the County Board of Supervisors. Dill argued that the only way to guarantee a fair election with these touch-screen systems was to provide them with a paper backup that voters could verify at the time they voted and that would be held as an audit trail by the county registrar. Dill provided the supervisors with a petition signed by hundreds of computer scientists from around the country in support of his position. Santa Clara County held off on the purchase until the vendor agreed to add at no cost a voter-verified paper ballot backup system.

As a result of Dill's efforts, the California Secretary of State Kevin Shelley created the Ad Hoc Touch Screen Task Force in February 2003 to look into the security concerns of these direct recording electronic voting systems. The committee reported its findings in July 2003. It called for enhanced security of voting systems through heightened federal, state, and local standards, certifications processes, and other procedures; through tighter security on distribution of software; and background checks on programmers working for the voting machine vendors. It also called for a permanent paper record of each voter's ballot, with the paper record taking precedence over the electronic record in most cases.

The Santa Clara County decision received national attention, and it represents the current thinking about e-voting generally. Internet voting, which is seen as even more risky, is not receiving serious attention for now, except for small pilot tests by the Department of Defense. Pleased with the results from the Voting Over the Internet pilot project in the November 2000 election, Congress mandated that the Department of Defense Federal Voting Assistance Program continue its pilot project of Internet voting and registration in the 2004 primaries and federal election. Fourteen states were invited to participate in the new project, known as SERVE (Secure Electronic Registration and Voting Experiment). Ten have so far expressed an interest. Members of the armed services, their families, and civilian personnel overseas who are registered to vote in these ten states will be able to vote, provided they have a Windows-based computer and Internet access. As many as 200,000 people could potentially vote through SERVE in the 2004 elections.

The Department of Defense contracted with Accenture's new eDemocracy business unit to provide a secure way to vote over the Internet. The security provisions are based primarily on the use of digital signatures, which involve the use of public key encryption to verify the identity of voters. SERVE has been criticized by Rebecca Mercuri and others, who note that the system put in place by Accenture provides security for the vote while it is being carried through the Internet, but that there is no assurance that the votes will be counted in an accurate way once they reach their destination. The Department of Defense will report on the experiment to Congress in 2005.

Although SERVE is the only experiment with Internet voting in the United States, a number of other countries are experimenting with and planning on implementing Internet voting later in this decade. The European Union began the Cybervote Project in October 2000, to test an Internet voting system that is verifiable and guarantees voter privacy. In the system, voters can cast their ballots from their personal computers, personal digital assistants, or mobile phones. The system is being tested in one Swedish, French, and German city in 2003. In April 2002, the Millenium Democratic Party, the ruling party in Korea, held its presidential primary online. In May 2002, the United Kingdom conducted 30 voting experiments to identify ways of enhancing voter turnout. Five local municipalities allowed Internet voting, and over 9,000 voted in that way. The highest penetration of voting by Internet (approximately 30 percent) came in two voting districts in Sheffield. In January 2003, the village of Anieres, a suburb of Geneva, was the first Swiss locale to allow Internet voting. Switzerland has a form of direct democracy, which typically requires local elections four to five times each year. In Anieres, 323 people voted by Internet, 370 by mail, and 48 went to the polls. The citizens voted in favor of a $3 million allocation of public funds to renovate a municipal property for use by a restaurant.

This section has focused so far on electronic and Internet voting, but there are some other election issues that involve the Internet and have received political attention. One issue involves the campaign finance reform law sponsored by Senators John McCain (R–AZ) and Russ Feingold (D–WI), signed into law by President Bush in 2002, and effective immediately after the November 2002 mid-term elections. The law prevents use of "soft" money contributions to pay for ads in favor of or against specific candidates for federal office, and bars many groups from airing ads that identify specific candidates by name within 30 days of a primary election or 60 days of a general election. Following passage of the act, the Federal Elections Commission (FEC) proposed that the law should apply to broadcast, cable, and satellite services, but not to text-based wireless ads, Internet ads, or web broadcasts. There is wide consensus that the campaign reform bill should not apply to private email or conventional web sites, but there has been great concern expressed by Common Cause and others that the law should apply to the functional equivalents of broadcast, such as Microsoft's interactive WebTV. The FEC ruled the Internet to be exempt from soft money restrictions. The constitutionality of the McCain-Feingold legislation will be decided by the Supreme Court in September 2003.

Another issue concerns congressional franking privileges and their online equivalent. It has long been the practice that members of Congress cannot send out mass mailings to constituents in the 60 days preceding an election. In 1996, when Congress adopted rules about the Internet, the House and Senate adopted different rules about changes to their government web sites prior to an election. Incumbent Senators are not allowed to make changes to their web sites during the last 60 days before an election, but there is no such restriction on members of the House. In the 2002 mid-term elections, there were four races in which members of the House were running against incumbent Senators, giving this slight advantage to the House member. The Senate plans to review the rule in 2003.

There are also examples of political cybersquatting. Jack Rollison (R–VA, Prince William County), a 17-year incumbent in the Virginia House found that a political supporter of his opponent in the primary election, Jeff Frederick, had registered *www.jackrollison.com* and was redirecting mail to the domain name to a Frederick site. Rollison complained, and the Frederick campaign released the domain name. Fredrick later accused Rollison of a copyright infringement by downloading unflattering photographs of Frederick from Frederick's personal web pages and using these photos in campaign ads. Rollison lost the primary to Frederick, but it is not clear how important the domain name issue was. More important seemed to be voter unhappiness over Rollison's support of a referendum to raise taxes in northern Virginia to fight traffic congestion.

Vices and Community Standards: Internet Pornography and Cyberpredators

Sex sells. A *New York Times* article estimated that there were 60,000 pornographic websites in 2000, triple the number only two years earlier. Already by 1998, pornography was estimated by CNET to be a billion dollar business and to represent 11 percent of all e-commerce. Today, Internet pornography is a much larger business than it was five years ago. Since 1996, Congress has repeatedly tried to legislatively control Internet pornography, but many of its laws or their key provisions have been struck down by the courts as unconstitutional.

The First Amendment of the Constitution protects freedom of speech from legal abridgement. There are two exceptions, however. One is obscenity. A famous court case of 1973, *Miller v. California*, established a three-part test of when a pornographic work is obscene:

1. whether the 'average person applying contemporary community standards' would find that the work, taken as a whole, appeals to the prurient interests;

2. whether the work depicts or describes, in a patently offensive way, sexual conduct specifically defined by the applicable state law; and

3. whether the work, taken as a whole, lacks serious literary, artistic, political, or scientific value.

The other exception is child pornography. Child pornography was defined by the courts in *New York v. Ferber* (1982) as material that "*visually* depicts sexual conduct by children below a specified age." Child pornography does not have to be obscene under the Miller test not to be protected under the First Amendment. This is because, according to *Ferber*, child pornography "is intrinsically related to the sexual abuse of children..." It has been child pornography that is the focus of almost all congressional action on Internet pornography.

In 1998 Ernest Allen, the CEO of the National Center for Missing & Exploited Children (NCMEC), and Louis Freeh, the director of the FBI, testified before a Senate committee about risks to children on the Internet. In response, Congress directed NCMEC to conduct the first scientific study of this problem, and the survey results (of 1501 U.S. children aged 10 to 17) from 1999 still provide the most accurate portrait we have of the problem. The study considered three kinds of risks to children: receiving sexual solicitations they did not want, viewing sexual materials they did not seek, and receiving threats or harassment. The survey showed that one in five children received a sexual solicitation over the Internet, and one in 33 received an aggressive sexual solicitation such as someone who sent gifts or asked them to meet physically. One quarter of the children had an unwanted online exposure within the past year to naked people or people having sex. One in 17 was threatened or harassed. Less than 10 percent of these incidents were reported to authorities, and less than half of these incidents were reported to parents. About one-quarter of the children report distress from these incidents. Older children (those over 14) in the survey were significantly more likely than younger ones in the survey to have experienced these incidents.

The popular image of the dirty old man as the perpetrator was not confirmed by the study. Many of the perpetrators were young adults and other children, and a large minority were women. This was even true of aggressive solicitors.

One major difference between Internet pornography and more traditional distribution forms is how easy it is for children to come across pornography unintentionally online. Searches for popular children's toys and games, including Beanie Babies, Barbie, and Candyland all have led children to pornographic websites. (Hasbro Toys has won back the Candyland.com domain name from the pornographers.)

Even though the easy access children have to pornography has angered both politicians and parents, the hard psychological evidence concerning the harm to children has not been clearly established. Researchers from the highly regarded Crimes Against Children Research Center at the University of New Hampshire have found:

> In the absence of evidence about the negative psychological effects of children's exposure to general pornography that could be used to justify regulation, antipornography activists have tended to cite other research about pornography; that it is used by child molesters in the seduction of children and that its consumption is sometimes a factor in the developmental histories of the child molesters themselves…. But unfortunately, despite its plausibility from anecdotal accounts, there is little research confirming a regular or causal role for pornography in child molestation. That is, it has not been shown that pornography results in the abuse of children who would not have otherwise been abused or the creation of molesters who would not have otherwise molested…. The harm-to-children issue is really about whether exposure to sexual materials causes psychological, moral, or developmental harm to children as a result of the viewing, and this is an eminently empirical issue on which virtually no research has been done. (Kimberly J. Mitchell, David Finkelhor, and Janis Wolak, "The Exposure of Children to Unwanted Sexual Material on the Internet: A National Survey of Risk, Impact, and Prevention," *Youth and Society* vol. 34 (March 2003), pp. 330–358, quoted from p. 334).

One final general point before turning to the history of legislative action: a March 2003 report from the U.S. General Accounting Office identified the different Internet technologies used to provide access to child pornography. They included the World Wide Web, Usenet, email, instant messaging, chat, and internal relay chat rooms, and peer-to-peer file sharing systems such as KaZaA. Particular concern was expressed about the ready accessibility of child pornography on the peer-to-peer networks because of their wide use by children to download music and video materials. Two independent test searches on selected keywords found that more than 40 percent of the downloads contained child pornography. Even searches on presumably innocuous keywords such as specific cartoon characters or celebrities yielded large percentages of child pornography.

Concern about pornography online began to mount in the United States in 1995. During 1993 and 1994 Senator James Exon (D–NE) had been trying to interest his colleagues in legislation on child pornography—to no avail. In June 1995, the anti-pornography advocacy group Enough is Enough provided Exon with a blue binder, which became known as "the blue book," with descriptions of pornographic material available free of charge on the Internet. Later, downloaded images were added to the book. The blue book was passed around on the floor of the Senate, and it was an impetus for the first congressional legislation in this area. Public opinion also began to get pumped up. That July, *Time* magazine published an article entitled "On a Screen Near You: Cyberporn—It's Popular, Pervasive, and Surprisingly Perverse."

The result was the quick passage, with overwhelming bipartisan support in both the House and Senate, of the Communications Decency Act of 1996. Essentially the bill prohibited indecent communications to minors by telephone, fax, or email, as well as displaying indecent material on the Internet that could be accessed by people 18 years old or younger. This law extended existing laws on obscenity and child pornography in several ways. Section 502(a), known as the "indecent transmission" provision, protected against "the knowing transmission of obscene or indecent messages to any recipient under 18 years of age." Section 502(d), known as the "patently offensive display" provision, protected against "the knowing sending or displaying of patently offensive messages in a manner that is available to a person under 18 years of age."

The American Civil Liberties Union, together with a number of other organizations, immediately upon passage challenged the constitutionality of the law on First Amendment freedom of speech grounds in federal court in Philadelphia. A three-judge panel decided unanimously to enjoin enforcement of these two key provisions concerning indecent transmission and patently offensive display. (*ACLU v. Reno, known often as Reno I*). The government filed an appeal, which went directly to the Supreme Court for review. The Supreme Court ruled in 1997 unanimously against the government. (*ACLU v. Reno, known often as Reno II*). The court found that the speech restrictions in CDA were overly broad and would have a chilling effect on free speech by adults. CDA raised broad but ambiguous restrictions on what speech was permissible for minors, and in order not to act unlawfully, adults on the Internet, since they had no good way of restricting their speech to adult recipients, would have to modify their speech so that it would always be suitable for minors. As the Supreme Court quoted another ruling in its findings: "the level of discourse reaching a mailbox simply cannot be limited to that which would be suitable for a sandbox." An adult-only audience, the court argued, could not be guaranteed in such an anonymous communication medium as the Internet, and any attempt to guarantee an adult-only audience, such as an adult-verification-number system or credit-card-verification system would impose too great a burden on non-commercial users of the Internet. There was also concern about the community standards provision in the CDA. Since the Internet cut across every geographical region, the community standard for whether a given communication is pornographic could be judged by the most conservative community, which went directly against the existing precedent from *Miller v. California* in which the test was local community standards.

The finding of the Supreme Court could not have surprised the legislators very much. Assistant Attorney General had privately advised Senator Patrick Leahy (D–VT) before passage that the law might unconstitutionally restrict free speech. Speaker of the House Newt Gingrich had spoken publicly about the constitutional problems at the time the bill was under consideration. The *New York Times* reported that President Clinton and his advisers were aware of the problems when he signed the bill. In fact, the bill itself contained a fast-tracking provision to get it reviewed by the Supreme Court in the event that it was challenged in court. The reason that the legislative and executive branch would knowingly go ahead with a flawed bill is pure politics. Nobody wanted to face an opponent in the next election who could ask why they had voted in favor of child pornography. One provision of the CDA that did stand the court appeals was the extension of interstate commerce in obscenity or child pornography to include interactive computer service, including the use of the Internet.

Congress responded to the unconstitutionality of the Communications Decency Act by attempting to make small changes that narrowly addressed the issues raised by the courts. The result was the Child Online Protection Act (COPA), sponsored by Senator Dan Coats (R–IN). COPA differed from CDA in three respects: it addressed only the World Wide Web and not the entire Internet; it applied only to commercial websites; and it narrowed the speech that was regulated from indecent material to "material that is harmful to minors."

COPA was supposed to have gone into effect into effect in November 1998, but it was immediately challenged in federal district court in Philadelphia by the ACLU and others. The district court enjoined enforcement of the law because of its likely unconstitutionality on First Amendment grounds. The district court found many things to object to in COPA. The cost of credit-card or age-verification systems by commercial web publishers, which were ways of protecting themselves from being charged under the law, were seen as economically burdensome and likely to cause these publishers to either not publish or self-censor the content so that they did not contain material harmful to minors. The district court noted that the law did not apply to foreign websites, which might contain material harmful to minors; that some minors had credit cards, which would undermine the credit-card system for restricting access to adults; and that other solutions, such as filtering systems, were a less burdensome remedy.

The decision of the district court was upheld in appellate court. The appellate court based its decision not on the same grounds as the district court but instead upon the community standards issue. Like the Supreme Court in CDA, the appellate court argued that a web publisher cannot restrict access of its materials to any specified set of locales, so it would be subject to the community standards of all communities, and hence to those of the most conservative community, contrary to the *Miller* test.

In May 2002 the Supreme Court ruled on COPA (*Ashcroft v. ACLU*). The ruling was narrow, stating "only that COPA's reliance on community standards to identify 'material that is harmful to minors' does not *by itself* render the statute substantially overbroad for purposes of the First Amendment." The Supreme Court remanded the case to the appellate court, which upheld the district court's ruling that the law is unconstitutional.

Even before CDA had been ruled unconstitutional, Congress had enacted another piece of child pornography legislation, the Child Pornography Prevention Act of 1996 (CPPA). The basic idea of this legislation was to expand the definition of child pornography to include visual depictions of a minor engaged in sexual activity, even if no actual minor is involved. This would include, for example, visual materials in which adult actors pretended to be minors, or computer images or paintings that represented minors engaged in sexual acts in the creation of which child models were not used. CPPA can be regarded as a follow-on to 1994 legislation that extended the definition of child pornography to include video that not only showed nude minors or that show the outlines of genitalia of minors through sheer clothing, but also to include video that focused on the pubic areas of minor females even though those areas were covered by opaque clothing.

CPPA was challenged in court, and in April 2002 the Supreme Court ruled that key provisions of CPPA are unconstitutional. (*Ashcroft v. Free Speech Coalition*) In particular, CPPA had broadened the definition of child pornography with the following definition:

"child pornography" means any visual depiction, including any photograph, film, video, picture, or computer or computer-generated image or picture, whether made or produced by electronic, mechanical, or other means, of sexually explicit conduct, where—

(A) the production of such visual depiction involves the use of a minor engaging in sexually explicit conduct;

(B) such visual depiction is, or appears to be, of a minor engaging in sexually explicit conduct;

(C) such visual depiction has been created, adapted, or modified to appear that an identifiable minor is engaging in sexually explicit conduct; or

(D) such visual depiction is advertised, promoted, presented, described, or distributed in such a manner that conveys the impression that the material is or contains a visual depiction of a minor engaging in sexually explicit conduct.

The court ruled that, for speech to be unprotected by the First Amendment, it must be either obscene or depict an actual child involved in sexual activity. Thus, paragraphs (A) and (C) in CPPA's definition of child pornography are acceptable, but paragraphs (B) and (D) are not. The court argued that laws that prohibited the use of real children in child pornography were constitutional because they focused on the production of the work, whereas CPPA focused instead on the content of the work.

The government unsuccessfully tried two other arguments in its defense of CPPA. First it argued that pedophiles could use child pornography produced without an actual child to encourage actual children's involvement in sexual activity, whet their own appetites, and thus to increase the likelihood of sexual exploitation of actual children. The court responded that the government "cannot constitutionally premise legislation on the desirability of controlling a person's private thoughts… The government may not prohibit speech because it increases the chance an unlawful act will be committed at some indefinite future time." The government's second argument was that, as imaging technology improves, it becomes harder to tell whether an image was produced using actual children or not; and hence harder to prosecute pornographers who do use actual children. The court was unmoved by this argument, noting that the "government may not suppress lawful speech as a means to suppress unlawful speech."

As in the case of CDA and COPA, the response of Congress to the unconstitutionality of CPPA was to try, try again. The most recent legislation is the Prosecutorial Remedies and Other Tools to End the Exploitation of Children Act of 2003, known generally as the PROTECT Act. The best-known feature of the PROTECT Act is its role to strengthen and nationally coordinate the Amber Alert systems that are in place in most states. The system allows massive and rapid dissemination of alerts over radio, television, and the Internet, when children are missing or abducted. The system is named after Amber Hagerman, a nine-year old girl who was kidnapped and murdered in Texas in 1996.

The nationalization of the Amber Alert system appears not to be problematic, according to legal scholars; but two other, less-well-known provisions of the PROTECT Act are expected to be disputed on First Amendment grounds. One provision makes it illegal to use a "misleading domain name" to lure people into viewing obscenity or luring minors into viewing material that is harmful to them. However, ambiguity over what makes a domain name "misleading" and over the definition of "harmful to minors" may lead the courts to throw this law out. The PROTECT Act also tries to get around the problems of CPPA and legislate against virtual child pornography. Whereas CPPA had broadened the definition of child pornography to include images that appear to be minors, the PROTECT Act narrows slightly the definition to "a digital image, computer image, or computer-generated image that is, or is indistinguishable from, that of a minor engaging in sexually explicit conduct." The writers of the PROTECT Act also have tried to get around the Constitutional problems of CPPA by adding exceptions for works with "serious literary, artistic, political, or scientific value." Legal scholars doubt that these changes will be enough to allow the virtual child pornography provisions in the PROTECT Act to pass constitutional muster.

There is one last piece of major federal legislation to consider, the Children's Internet Protection Act of 2000 (CIPA), which required public schools and libraries to install software on their Internet computers to filter out pornographic sites if they want to certain kinds of federal funding. Prior to the passage of CIPA, according to a survey by the Library Research Center, while most libraries had an Internet use policy in place, only 17 percent used filters on some their computers and only 7 percent used filters on all of their computers. The most common alternative protection measure, adopted by about two-thirds of libraries, was to require parental permission for their children to use the library computers to search the Internet.

In a landmark case, in 1997 the Board of Trustees of the Loudoun County Public Library in Virginia required that the six public libraries install X-Stop filtering software on all computers to block access to sites that provided child pornography or obscenity, as well as sites that were deemed "harmful to minors" under Virginia statutes and case law. A group of Loudoun County residents (under the name *Mainstream Loudoun*) sued that X-Stream blocked access to sites that were protected under the First Amendment, the policies set forth for blocking by the Board of Trustees were unclear, and enactment of this filtering had chilling effect on receipt of constitutionally protected materials by library patrons. In 1998 the federal District Court for the Eastern District of Virginia ruled in favor of the residents. The judge admitted the arguments of the Board of Trustees that there was compelling reason to minimize access to illegal pornography in the libraries and avoid a sexually hostile environment. However, the judge also ruled that the filtering practices adopted by the Board of Trustees were neither necessary to further those interests nor narrowly tailored to do so. The judge identified less restrictive alternatives, such as filtering only on Internet terminals for minors, more diligent monitoring by libraries, or privacy screens around terminals.

The *Mainstream Loudoun* decision rallied support for the School Filtering Act being promoted by Senator John McCain (R–AZ), requiring libraries to install filters on their Internet terminals if they wanted to receive two important sources of federal support: the universal service (also known as "e-rate") discounts to libraries provided by the Communications Act of 1934 and LSTA Funds provided from the Library services and Technology Act of 1996. Congress passed McCain's bill into law as the Children's Internet Protection Act of 2000, as part of the Fiscal Year 2001 Appropriations Bill. The law did include a disabling provision, by which a library authority could disable the filtering for a patron for research or other lawful purpose.

In March 2001, a month before CIPA was to go into effect, it was challenged by lawsuits from both the ACLU and the American Library Association (ALA). These two suits were consolidated, with the ALA taking the lead role. Much of the concern was over the inadequacy of the filters. Although the filters under-filtered (allowed access to some pornographic websites), the constitutional concern was that the filters over-filtered (blocked access to web sites that were protected under the First Amendment). Other concerns were raised over community standards being imposed at a national level (similar to the problems with CDA and COPA), that the filters must be installed on all machines so that adult patrons must access terminals set to the standards for children, and that the guidelines for disabling were vague and may have a chilling effect on adult use.

In May 2002, a three-judge panel representing the federal District Court in Philadelphia ruled unanimously that CIPA violated the First Amendment by blocking access to substantial amounts of protected speech and is thus unconstitutional. Similar to the Loudoun case, the court noted that libraries have a variety of less restrictive alternatives than filtering every Internet terminal: offering filtering software as a choice for Internet searching at the library by children, education and Internet training courses, enforcement of existing Internet use policies by library staff, and placement of terminals and use of privacy screens to limit minors from viewing adult searches.

In a surprising turn of events, in June 2003 the Supreme Court overturned the decision of the District Court and allowed the government to require filtering software as a condition of receiving the public funds mentioned above. Four of the justices found the law constitutional, and two more found it constitutional so long as libraries would turn off the filters upon the request of adult library patrons, and that patrons would not be required to explain why they made this request. Thus the filters are being seen by some legal scholars as a kind of opt-out mechanism.

Before leaving this topic, it is useful to look at the tests of filtering, to see the extent to what degree they under-filter and over-filter. There have been a number of empirical studies, but the two that have received the most attention are one by *Consumer Reports* in March 2001 and one by the Kaiser Family Foundation in December 2002. The *Consumer Reports* study had fairly pessimistic results. It found that five of six of six commonly used filtering programs did not block more than 20 percent of the objectionable sites. One of them (AOL Young Teen) blocked 86 percent of the objectionable sites, but it also seriously over-filtered

(63 percent of legitimate sites). The Kaiser report had results that were interpreted as negative by the ALA and as positive by the *Journal of the American Medical Association*. The Kaiser report showed that, when the filtering software was set at its intermediately permissive setting, the software on average caught 90 percent of the objectionable sites while blocking only 5 percent of non-objectionable health sites. When the software was set at the least permissive setting, blocking of objectionable sites only increased by 1 percent (to 91 percent), but over-blocking of health sites increased dramatically (to 24 percent). Online health information is widely regarded as a good test of over-blocking because 70 percent of 15 to 17 year olds say they have used the Internet to look up health information, yet health sites have information on topics that might easily be blocked, such as safe sex practices, sexually transmitted diseases, pregnancy prevention, abortion, and homosexuality.

Congress has also provided increasing levels of financial support for various federal law-enforcement efforts in the areas of child pornography and child predators, often carried out in collaboration with state and local agencies. The National Center for Missing & Exploited Children is supported by the Justice Department. Justice has also established some 30 Internet Crimes Against Children task forces around the country to investigate crimes and provide technical assistance and train personnel at the state and local levels. A new web portal, called Law Enforcement Data Exchange (LEDX), has been established for securely sharing sensitive information and protecting the chain of evidence for legal purposes in investigations. The Homeland Security Department has recently established Operation Predator, to coordinate related efforts across the various organizations that were pulled together into the Homeland Security Department. The effort is being led by the CyberSmuggling Center of Fairfax, Virginia, a unit of the Bureau of Immigration and Customs Enforcement. Other Homeland Security organizations involved in these efforts are the FBI, Justice Department, Postal Inspection Service, and the Secret Service.

Although the focus in this section so far has been on child pornography and child predators, the Internet has become a tool that is used increasingly often when predators target adults (usually, but not always women) in the form of cyberstalking. Stalking is harassing and threatening behavior that a perpetrator engages in repeatedly, such as following a person around or making threatening telephone calls. The Internet provides new opportunities for stalking. It is easy online to get personal information about the victim, including personal address, unlisted telephone numbers, and financial information. It is easy to send repeated harassing or threatening messages by email—even to automate this process. The perpetrator can easily conceal his identity on the Internet by using false names or anonymizers. It is easy to send out mass mailings that exhort or trick others into sending harassing mail.

Three examples illustrate the seriousness of cyberstalking, which anecdotally appears to be rising rapidly (some estimate about 100 cases per week currently).

In the first successful prosecution under the new California cyberstalking law, a security guard targeted a woman who rebuffed his romantic advances. The security guard impersonated the victim in various chat rooms and online bulletin boards, where he posted her telephone number and address, together with her supposed fantasies of being raped. On six occasions men came to her door, indicating they wanted to rape her. In another case, in New Hampshire, a stalker purchased the victim's personal information online from an information broker (Docusearch), set up a web site where he referred to plans to stalk and kill the victim, and fatally shot her several months later. In a recent case, a Colorado woman was mistakenly identified in several chat rooms as the women who was going to testify in the rape case against basketball star Kobe Bryant. She received hundreds of threatening messages online from basketball fans.

The first traditional stalking law was passed in 1990 (by California), and today all 50 states have such laws. Only about a third of the states have separate cyberstalking laws, though some additional states have laws against stalking via electronic communications. A shortcoming of many of these laws is that they focus on the electronic analog of traditional stalking and do not address new opportunities for stalking through the Internet, such as sending out notes to chat rooms that encourage third parties to take some harassing or threatening action on the victim. It is clear that the state laws are having difficulty keeping up with the pace of technological innovation. Jurisdictional issues arise when the stalker is in a different state or country than the victim.

The situation at the federal level is less advanced. The Cable Communications Policy Act of 1994 has become an increasing problem in cyberstalking investigations, as more people use cable modems to connect to the Internet, because the law prohibits cable companies from disclosing cable subscriber records to law enforcement agencies without a court order and advance notice to the cable subscriber. A number of cases referred to the FBI have been turned over by the U.S. Attorneys' Offices to state and local law enforcement because there were no grounds for federal prosecution. The Violence Against Women Act of 2000, sponsored in the Senate by Joseph Biden (D–DE) and Orrin Hatch (R–UT), mainly reauthorized various federal programs to fight traditional kinds of violence against women, including traditional stalking. It does not speak directly to online stalking, although it does authorize funds for training that can be used for training law enforcement agents about cyberstalking, and it specifically criminalizes interstate travel to engage in cyberstalking. Vice President Gore took an interest in cyberstalking and asked the Justice Department in 1999 to study the issue. A study was issued by the Attorney General that same year. An updated report, with very similar material was part of a report to Congress from the Justice Department in 2001. However, it is clear that this issue has not yet become an issue for coordinated and prioritized federal effort as of yet.

Vices and Community Standards: Internet Gambling

Gambling is a good example of an activity that has been highly and relatively effectively regulated in the past, for which the introduction of the Internet raises new regulatory problems. Internet gambling first appeared in the mid-1990s and now represents a $5 billion industry worldwide, with approximately 60 percent of the betting placed from the United States. However, the vast majority of the Internet gambling sites—about 1800 currently—are operated from outside the United States. In more than 50 countries, Internet gambling is legal. Thus serious jurisdictional and law enforcement issues arise.

Various criticisms are levied at Internet gambling. It is regarded as a particularly addictive form of gambling, and it provides an easy way for children to become involved. Consumer protections are weak. This form of gambling is also seen as a threat by law enforcement agencies as a potentially effective means for money laundering by organized crime, although the banking and credit card industries believe these concerns to be overblown.

Internet gambling is subject to both state and federal laws. For the most part, gambling is regulated at the state level—ranging from no gambling of any form permitted in Utah, to a large but heavily regulated industry in the neighboring state of Nevada. As for gambling on the Internet, most states ban it. One exception is online betting on horse races allowed in about a dozen states through a special arrangement between TVG, a division of Gemstar-TV Guide International, and the National Thoroughbred Racing Association. The 1978 Interstate Horseracing Act, as amended in 2000, allows this interstate betting, provided certain strict rules (all of them met by the GTV system) are obeyed.

Federal laws basically are intended to ensure that interstate and international commerce do not circumvent state laws. The core federal law is the Wire Act. Passed in 1962 to deal with telephone betting, it is being applied today to Internet gambling as well. The Act prohibits gambling businesses from knowingly placing bets or sending information that helps to place bets over interstate or international wires. Foreign businesses offering Internet gambling to U.S. citizens are not immune from this law, although enforcement may be problematic.

There are some questions about the scope of the Wire Act, which court decisions have not entirely clarified. The Act as written talks about "bets and wagers on any sporting event or contest," so there is some question as to whether it applies to forms of gambling not related to sporting events. There are also two questions of ambiguity in the act's language where it discusses "transmission of a wire communication." Does the Act apply to satellite and wireless communications? Does it apply in situations where betting information is only received, not transmitted by the business? The Act also is sometimes interpreted to give an exception to Internet gambling in cases where the gambling business and bettor are both located in jurisdictions in which betting on the event is legal, so long as the Internet transmission contains only information that assists in placing bets and does not contain the bet itself.

Legal scholars have suggested that two other federal laws could also be applied to Internet gambling, although neither has yet been the basis for a federal action. The Travel Act criminalizes interstate and foreign commerce to distribute the proceeds of unlawful activities. The Illegal Gambling Business Act makes it a federal crime if the business activity is illegal in the state in which the business operates, the business employs at least five people throughout a 30-day period, and either operates for more than 30 days or grosses at least $2,000 in a single day.

Credit card companies are wary of Internet gambling, and their actions have reduced the amount of this gambling—for if one can control the flow of funds, one can control the gambling. The wariness stems from concern that Internet gambling is of questionable legal status, and the credit card companies do not want to aid illegal activities; because Internet gambling is regarded as a high-risk industry, vulnerable to fraud and other illegal activities; and because there is also high risk of financial losses from gamblers who refuse to pay their gambling debts. The credit card associations may have reason to be wary on this last point. In one case, a credit card company sued an individual for not paying gambling debts charged against the credit card. The customer counter-sued, arguing that the liability resided with the credit card company because it allowed the customer to engage in an illegal activity with the card. There was some uncertainty about what the legal outcome would be, and the case was settled out of court.

The credit card companies that themselves manage both users and merchants, such as Discover and American Express, try to limit Internet gambling by refusing gambling organizations from becoming merchants authorized to accept their credit cards. The situation is more complicated with the credit card associations, such as Mastercard and Visa. Visa, for example, allows each individual member bank to determine whether it will accept Internet gambling transactions. Almost all U.S. banks do try to block Internet gambling transactions, but there are some overseas banks (and perhaps a few U.S. banks) that do not. The principal strategy of the credit card associations has been to provide a coding scheme that enables a bank to readily identify e-commerce transactions and another code for gambling transactions. An automated system can be programmed to reject those transactions that contain both the e-commerce and gambling markers, for banks that want to refuse Internet gambling.

As one might expect, the gambling businesses try to find ways around the coding scheme. Sometimes they deliberately miscode the transactions. If they have another business, they sometimes code their gambling transactions with the code numbers of their other businesses. Sometimes they engage in "factoring" by which the gambling organization has another merchant submit their gambling transactions (with a different transaction code) for a small per-transaction fee. The credit card companies review Internet gambling websites, place test bets to see if the gambling organizations are using their cards for gambling fees against credit card association guidelines, and do data analysis on patterns of transactions. So far, the credit card companies appear to be catching about 80 percent of Internet gambling transactions, but as with all technical solutions, there is constant innovation on both sides as each tries to get the upper hand.

Another scheme being used by the Internet gambling organizations is the use of electronic payment providers or payment aggregators to pay for Internet gambling activities. Pay Pal, for example, was offered as a payment means by three-quarters of Internet gambling websites, according to a General Accounting Office study in 2002. Others included Firepay, NETeller, EZPay, and Equifax. New forms of electronic payment (e-cash) are being developed all the time.

Because of the emergence of Internet gambling in the mid-1990s, Congress established the National Gambling Impact Study Commission in 1996. A 1999 report of the Commission urged Congress to pass legislation outlawing Internet gambling not already legalized and prohibiting use of credit cards for Internet gambling payments. In every session of Congress since 1998, bills have been introduced and in two cases have received Senate approval, but none so far have been enacted as law. Some of these bills outlawed Internet gambling outright. Others focused on closing loopholes in the Wire Act or in banning the use by Internet gambling businesses of payment by customers using checks, credit cards, and electronic funds transfers.

In June 2003 the House passed the Unlawful Internet Gambling Funding Prohibition Act, sponsored by Representative Spencer Bachus (R–AL) and co-sponsored by Michael Oxley (R–OH), chair of the House Financial Services Committee. The core of the bill focused on tightening the purse-strings—requiring regulations that identify and block gambling businesses from accepting checks, credit cards, and electronic funds transfers in connection with any illegal Internet gambling operation. Exempted from the bill are any state-licensed or state-authorized gambling practices. A companion bill has been introduced in the Senate by Jon Kyl (R–AZ), Richard Shelby (R–AL), and Dianne Feinstein (D–CA).

Lobbying on Bachus's bill led to some unusual alliances. The bill was supported by both moderate (United Methodist Church) and conservative (Christian Coalition, Family Research Council) church groups, and by the National Thoroughbred Racing Association. Opposed have been the Interactive Gaming Council, the conservative Traditional Values Coalition, and various Native American groups such as the National Indian Gaming Association. One can understand how the church groups would favor the legislation, but what about the others? The horse-racing community already has the only legalized online system and would prefer no competition. The gaming council prefers regulation (as used in Las Vegas) over prohibition. The Traditional Values Coalition opposes provisions in the House bill that allow continuation of casinos, dog and horse racing, and state lotteries that are already allowed by the states; and would prefer no law to one with these exceptions. The Native Americans want specific language in legislation that allows them to extend their rights to run casinos on tribal land so that they can offer online betting as part of their operations.

Two additional considerations complicate the picture. The Justice Department would prefer to have seamless legislation that would enable it to investigate both traditional and online gambling. Thus the Justice Department would prefer to have an extension of the Wire Act that clarifies issues such as wireless and satellite transmission than to have new legislation for Internet gambling separate from the Wire Act. In the debate on the Bachus bill, Judiciary chairman Jim Sensenbrenner (R–WI), John Conyers (D–MI), and Chris Cannon (R–UT) introduced an amendment to remove the exception that allowed state authorized Internet gambling. The amendment was defeated largely on the basis of state rights.

It is too soon to know what the outcome of this legislative initiative will be. There does not seem to be the same driving pressure to legislate Internet gambling as there is to legislate online pornography.

Further Readings

Mark C. Alexander, "The First Amendment and Problems of Political Viability: The Case of Internet Pornography," *Harvard Journal of Law and Public Policy* 25 (2002), pp. 977–1030.

Henry Cohen, "Child Pornography: Constitutional Principles and Federal Statutes," CRS Report for Congress, Congressional Research Service, Library of Congress, updated June 25, 2003.

Kevin Coleman, "Internet Voting," CRS Report for Congress, 2003, Congressional Research Service, Library of Congress, updated January 31.

Michael A. Geist, "Is There a There There? Toward Greater Certainty for Internet Jurisdiction," *Berkeley Technology Law Journal* (2001).

Rebecca Mercuri, "A Better Ballot Box?" *IEEE Spectrum* October 2002, pp. 46–50.

Annette Nellen, "Overview of E-Commerce Taxation—Guide to Understanding the Current Discussions and Debates," July 2003, College of Business, San Jose State University, *http://www.cob.sjsu.edu/facstaff/nellen_a.*

David F. Norden, "Filtering Out Protection: The Law, The Library, and Our Legacies," *Case Western Reserve Law Review* 53 (2003), pp. 767–814.

Nonna A. Noto, "Extending the Internet Tax Moratorium and Related Issues," Congressional Report Service, Library of Congress, updated January 17, 2002.

U.S. General Accounting Office, "Internet Gambling: An Overview of the Issues," GAO-03-89 (December 2002).

Jonathan Zittrain, "Be Careful What You Ask For: Reconciling a Global Internet and Local Law," Harvard Law School Public Law Research Paper No. 60 (2003).

***

Thanks to Indiana University graduate student Jason Gretencord
for his extensive and excellent research assistance on this chapter.

Chapter 5

Computer Security and Critical Infrastructures: A Fine Line between Secure and Vulnerable

David Bruggeman

"The Vatican has warned Catholic bishops and priests not to use the Internet to hear 'online confessions' in case they are read by 'ill-intentioned people such as hackers' for purposes such as blackmail."

—Richard Owen, The Times (London), January 10, 2003

Even the Vatican would agree there is nothing sacred where the Internet and computer security are concerned. As online activity increases and more computer networks operate in the workplace, government, and other parts of daily life, attacks on those networks will increase. More people are familiar with physical security than computer security. Following the 2001 terrorist attacks both kinds of security are getting more attention, and are more similar than you might think. When it comes to computer security, however, it's not as good as you think, it's as reliant on humans as on technology, and its consequences run counter to most of the perceived benefits of the Internet. Making computers and critical infrastructures (such as the banking network, the phone system, and the power grid) more secure requires finding a better balance between access and security.

Like all of the issues covered in this book, issues in computer security and critical infrastructure highlight differences over values and priorities. Resolving these differences are matters of policy and politics, regardless of the technology involved. After all, that technology will eventually be replaced. All of these differences require finding a balance between competing interests such as:

- Privacy versus protection,

- Protecting companies versus increasing knowledge,

- Interaction versus restriction, and

- Regulation of the private sector versus the stifling of innovation.

The major tension in this chapter is how to encourage or ensure a reasonable amount of security for critical infrastructures and computer networks without undercutting the robustness and responsiveness of these networks that make them so appealing in the first place. This requires navigating a rough landscape of doomsday scenarios, realistic threats, and scarce knowledge about what has happened, what could happen, and what will happen.

One of many consequences of the current state of security is the loss of trust. Trust is an essential requirement for online activity. Online retailers need to demonstrate the security of their transactions in order to gain the trust of potential customers. Some people will not make purchases online because they don't trust that their credit card or other personal information will be secure. A notable exception to this insistence on security is that computer companies do not receive the same level of scrutiny from customers for the security of their products and services. This topic is discussed later in this chapter.

Some people trust (perhaps without much thought) computer systems to function most of the time. Businesses and governments rely on them to do their work and interact with the world. Critical infrastructures rely on computers to manage essential services. They help keep the trains running and the lights on. Computer networks are an important part of operating these critical infrastructures, and keeping them secure is very important. While computer viruses have existed since the mid-1980s and attacks on critical infrastructures have taken place since the early 1960s, the pervasive nature of computers in the early twenty-first century requires that computer security be revisited and revised as each new service or technology (mobile phones, PDAs, instant messaging, wireless networks) is developed and implemented in society. Policies must be flexible enough to handle changes in technology while still possessing enough bite to be effective.

The risks of failed security can be much greater than a lost credit card number. Organizations can be paralyzed and lives placed in danger from a cyber attack. Until the Slammer worm spread through the Internet in early 2003, most viruses and worms were nuisances (most of those released still are just nuisances), belying the potential impact of such attacks. While no casualties were attributed to the Slammer, its impact on critical infrastructures put lives at potential risk. A computer virus that infiltrates a 911 system compromises the ability of emergency services to respond to a crisis. If significant losses (of life or property) result from the next virus release, or the nuisance from viruses like Blaster and Sobig increases, trust in computer networks will be further compromised. This could prompt customers, insurers, and the government to demand better security from the computer industry. But it could also prompt many consumers to stop using computer services. The best interests of computer-related or computer-dependent businesses suggest that security action be taken before lives are lost or consumer trust is broken.

While the possible outcomes are dire, they are still only possibilities. And like many issues related to information technology, some believe that those who emphasize the possible are overreacting. As those who follow the climate change debate may understand, it is difficult to justify preemptive action against a potential threat when the costs of such action are seen by some as a greater harm than doing nothing. Perhaps it will take a more significant failure of computer security to prompt greater action. Policy makers might consider whether the damage from such a failure is worth postponing action in computer and critical infrastructure security. They must also consider whether significant investments in security should be made if the improvement to overall security is marginal. If investments are required, a key policy question will be what kinds of investments. Should the money be spent on new technology, incentives for consumers and the private sector to change their behavior, or some combination? Risk and benefit assessment will be a critical component of a successful strategy for ensuring cybersecurity.

The Slammer worm will be the starting point for this chapter, since it helps illustrate many of the issues and tensions in computer security. Some historical and policy influences on the current computer security environment will be discussed, and issues concerning computer security will be analyzed in terms of tradeoffs and tensions. The Internet was a bottom-up development, emerging from a number of individual computers and networks. This helped give computer networks a robustness and responsiveness that is attractive and should be preserved. But in order to increase security, this robustness will need to be tempered. Competing interests must be negotiated, and choices must be made about how to temper this robustness.

Getting Slammed

The Slammer worm that attacked the Internet in late January 2003 is a good example for examining the challenges and consequences of computer security. This worm (a type of virus that replicates itself from system to system without need of a host file), also called the Sapphire or SQL worm, was the most severe attack on the Internet and connected computers since the Code Red and NIMDA viruses of 2001. The worm took advantage of a flaw in Microsoft SQL database software that was acknowledged by the company in July of 2002. While it may not have affected as many computers at the Code Red virus due in part to flaws in its design, Slammer is considered a more significant attack for several reasons. The number of infected machines doubled in size every 8.5 seconds when it first appeared, reaching its full rate of scanning for vulnerable systems after about three minutes. It affected systems up to three days after its initial release. Estimates of lost productivity from the worm (due primarily to slower Internet traffic) range from $950 million to $1.2 billion over the first five days following its release (which happened on a weekend), making it the ninth most costly malicious code at the time it was deployed. Slammer was only 376 bytes of computer code, and infected 90 percent of all vulnerable servers within ten minutes, making it a candidate for a Warhol worm, so named because it could have taken over the Internet within fifteen minutes.

Aside from the wider audience of the virus (it affected customers and other people outside of the companies infected by the worm), the Slammer worm and the recovery from it highlight the uneven response to viruses by those involved in computers. A patch to correct the flaw in the Microsoft SQL software was available in July of 2002, but several systems did not have the patch installed. While sometimes patches can be more trouble than they are worth, they are the traditional response to security flaws and do the job they were designed to do. Slammer utilized an older, less secure protocol, UDP, to spread through the Internet. This contributed to its rapid spread (along with being able to stay in RAM memory). Slammer took advantage of insecure software, uninstalled patches, and an older, less secure protocol. A system that had any one of those characteristics was vulnerable, but each of them can be fixed.

The worm also affected critical infrastructures. Parts of the Bank of America's ATM network went down, and a nuclear plant in Ohio was infected as well (although the plant was already offline). More seriously, dispatchers at the 911 telephone system in Bellevue, Washington had to use manual systems when their computers were affected. Lives could have been lost as a result. This demonstrates that such supposedly self-contained systems like 911 networks are indeed vulnerable to Internet viruses, in part because they are not the stand-alone systems they used to be. If such systems are susceptible to worms, they are also susceptible to more directed attacks. As many critical infrastructures are privately owned, the government will not be able to effectively protect these systems without better monitoring, regulation, and cooperation of public and private networks.

Whoever developed the Slammer worm likely had some indirect assistance. A "proof of concept" code circulated publicly in the summer of 2002. Such a code demonstrates the vulnerability in the program being tested, and prompted the development of the patch for the flaw. While it is important for software manufacturers to know that their product has bugs or other vulnerabilities, public disclosure opens the software for the possibility of exploitation. In this situation, the competing interests are the public's right to know the security of the products they rely on, and the potential harm to businesses and the possible dampening of innovative pursuits for fear of exploitation by hackers or competitors.

Microsoft received a fair amount of criticism in connection with the Slammer worm, and a class action lawsuit may proceed in South Korea, which was significantly affected by the worm. Part of this criticism has to do with timing—the company was rolling out its Trustworthy Computing initiative around the time Slammer hit. Also, Microsoft is a preferred target of consumers and some in the computer industry because of its status as a major player and because its actions in light of (and those that prompted) recent antitrust litigation. But Microsoft, like most software providers, was doing what was legally required and expected of them. Once Microsoft knew about the flaw, they devised a patch and made it available to consumers. The burden of computer security currently rests with both commercial and home consumers. Our dependence on computers makes it difficult for simple market pressures to prompt industry change, especially when a few firms dominate the market. To some extent there is a captive market. There are no product liability provisions that encourage computer developers to increase the security of their products and services. Perhaps it's important, given the contributions of information technology to the economy, to restrict any potential lawsuits. However, increased expectations of product quality could provide for better products, as well as a decrease in lost productivity due to worms, viruses, and other security trouble.

Security of computer systems and critical infrastructures is an important, but underappreciated, aspect of the national and global computer landscape. If people are expected to rely on computer systems and critical infrastructure to the extent that modern life demands, they should feel secure in doing so. The issues highlighted by the Slammer worm suggest that several practices, products, and services need to be revisited and discussed as computers and their networks continue to permeate society and technology continues to change. While technical issues are a part of successful security, computer and critical infrastructure security involves a great deal of social engineering as well. Policy is a form of social engineering, and many of the solutions to problems demonstrated by the Slammer and other worms will involve changes in behavior that are required independent of the technology involved.

The environment in which the Internet and networked computing developed has changed dramatically. The assumptions, protocols, and policies of computer and critical infrastructure security need to better reflect the changing landscape. In an age where policies and politics are even more behind changes in science and technology, it is even more important that policy solutions are not tightly coupled to specific technologies.

Background

Dial-up connections to the Internet might be on the way out, but the telephone system is still an influence on the Internet and other computer networks. This is due in part to the networked nature of the telephone system and in part because it serves as a regulatory model. Instead of starting completely from scratch when developing rules and laws for computer systems, many legislators look to the phone system as a guide, if not a model, for developing regulations for the Internet.

It's important to note the similarities and differences between the two kinds of networks. The switching systems of the two networks are different (though telephone switching is more digital than before). Computers are becoming less reliant on phone lines for their connections and migrating to cable systems, wireless operation, and other forms of signal transmission (though the telephone network is not as land-locked as it used to be). With changes in technology likely, future policy solutions and processes should take care not to be too invested in particular technologies as the basis for their decisions. The focus should be on the services that are provided by information technology, independent of specific technologies. Policies are tough to revise, and making them independent of specific technologies allows for a more flexible response to changes in the technology and/or the policy environment. But given current technology changes in telephones, it is still reasonable to keep an eye on changes in both telephone technology and telephone regulation as it could influence regulation of computers and other networked devices.

Another historical influence on computer security is cryptography. Utilized primarily by the military and the government to secure communications, computers revolutionized the field. They allowed for codes to become longer and more complex. Code breaking can be at least partially automated thanks to computers. There is a mindset associated with the early military uses of cryptography that lingers still—"total security." It is similar to a "fortress" mentality that can be seen in some security systems. Such a goal conflicts with a major advantage and intent of computer networks—interaction and interconnection. While total computer security for a non-networked system is difficult; total security for a publicly networked computer is even harder. Security measures slow down the speed of communication as incoming and outgoing traffic is checked for viruses, the sender and receiver are authenticated, and other measures are implemented. Security measures make it harder to access a particular computer/business/website, but a networked world prizes access. This tension is the source for many dilemmas (both technical and policy) involving computers and security.

A step taken by many systems to ensure greater security is to disconnect from larger networks like the Internet. The military's secure networks are disconnected from the outside world. Such measures are most effective when points of access are well controlled. A person could take a laptop home, connect to the Internet, infect that machine and then infect the secure, disconnected system once they return to work simply because they weren't careful enough. Security relies on people at least as much as it relies on technology.

Security was not a top priority when the Internet was first developed. While the system was developed in part to ensure communication in times of national emergency, the Internet is primarily a product of academic interests. Their major goal was to share information, and sharing information is not always consistent with securing an environment. The academic environment encourages the free flow of information, and is not going to be as concerned about whether people should be accessing that information. The Internet also started out small. That it might go very public and become nearly ubiquitous may have been a dream, but scalability was not a consideration in its initial design. Nor were other future possibilities such as the consequences of coding years in two digits rather than four when the centuries changed. It is impossible to effectively predict all possible consequences of a design. Total security, or perfect computing, is a goal, one that can never be achieved. As Joe Hartman, director of North American antivirus research at TrendMicro said, "We have recognized over the last several years that you cannot prevent a virus. There will always be an entry point."[1] You cannot design a perfect system. Even if you could, needs and capabilities do change over time; what might have been perfect in 1985 is not likely perfect in 2004. Total security might be a laudable goal, but anyone who promises it is fooling themselves and their customers. That customers, and lawmakers, might expect total security demonstrates a lack of understanding or an unwillingness to approach security from a network perspective.

The Internet and computer systems are open to those who seek to operate anonymously. When the Internet first became widely available, a person could really be anyone or anything. In most cases, there was little a person could do to authenticate the identity of the person that they were talking with. A classic example is the attractive 19-year-old coed in the chat room who is really a 54-year-old man. The lack of authentication in computer systems is a significant

---

[1]   Lemos, Robert. "Slammer report: More headaches," *ZDNet News*, February 7, 2003. *http://zdnet.com.com/2100-1105-98376.html*, accessed February 8, 2003.

security concern. Sensitive information should only be available to a limited number of people, and you need to be able to authenticate someone who tries to gain access to the information to limit that availability.

The Y2K bug (in which the year 2000 rolled around, problems might result if data with two-digit years was read as 1900) was perhaps the first major demonstration of how computer practices and processes could have serious unintentional consequences. While the doomsday scenarios predicted by some failed to materialize, Y2K showed that even the smallest choices can have profound consequences. It also demonstrated the extent to which critical infrastructures (systems like the power grid, financial networks, transportation systems, etc.) are dependent on computers in order to function effectively. Given the lack of incidents concerning Y2K, some considered it a false alarm and the efforts to beat it overreactions. Many make a similar argument about the possibility of cyberattacks or cyberwarfare. More realistically what happened with Y2K was like dodging a bullet. Such dodges have happened with many recent viruses. While they have been disruptive, in many cases some flaw in the programming or execution of these viruses makes them less of a problem than they could be. There is potential for significant damage to be done, but the fact that it has not happened yet makes some think it will never happen. A more prepared policy approach would take some place in between that extreme and the opposite possibility (and paranoia) associated with an electronic "Pearl Harbor."

A computer component doesn't have to be a critical system or function in order to be a security concern. The interconnected nature of computer systems allows for a virus, or other attack, to hit a "low-risk" area, and through interconnections between systems, or a simple denial of service attack, infect and/or disable the "high-risk" parts of the system. The Slammer worm attacked database systems, which were not identified as a critical system. This is important because some federal information security legislation places particular emphasis on securing critical systems. Slammer demonstrated that a non-critical system that allows access to a critical system can be as much of a risk as an unprotected critical system. While you can reduce such interdependencies and increase security, it typically involves a tradeoff with the robustness of a system.

An important part of Y2K was the government support of industry in responding to the potential crisis. Part of their efforts included passing both federal and state legislation to encourage the sharing of information and limiting (but not eliminating) the liability for problems related to Y2K. The relevant federal legislation was the Year 2000 Information and Readiness Disclosure Act. The federal government did not effectively encourage a similar sharing of information regarding computer and critical infrastructure security incidents following the attacks of September 2001.

Domestic security in general became more important to the United States government following the attacks on New York and Washington in September 2001. Cybersecurity is part of the increased American interest in domestic security. While the Internet coped with the physical attacks, the attacks demonstrated the interconnectedness of computer systems as well as how they don't always correspond to physical geography. The Internet may be global, but its infrastructure has defined paths through which data travels. Important facilities in Manhattan were knocked out, and online services in several countries were affected. Lower Manhattan serves as a hub for Internet traffic, and is a key access point for several underseas data cables as well as servers involved in coordinating Internet activity. South African websites were adversely affected because the Domain Name System server was located in Manhattan, and many South African sites were due for renewal shortly after the attacks. While the Internet is robust and can reroute traffic around blocked or broken connections, it requires multiple pathways to do that. Intercontinental connections are relatively few because they rely on transoceanic fiber, so the network is not as robust between continents as it is within them. As demonstrated by power outages, failures in a critical node or junction of the Internet could dramatically affect services far away from the disruption.

While computer networks like the Internet are relatively young, they have a history, and are influenced by technologies that preceded them. Information technology is reliant on communications technology for data transmission, and they share many technical characteristics. The telephone serves as a regulatory model, a model that is becoming less and less appropriate as computer technology diverges from the operating characteristics of the telephone system. The environment in which the Internet is used is much more active and vulnerable than the environment it was created and nurtured in. Relevant policies must be flexible enough to deal with the new environment. Computer networks enjoy a certain degree of flexibility and robustness because they are networks, but they present a series of risks as well. Security processes must adapt to meet the risk and maintain flexibility.

Legislation and Policy

There are several laws and national policies that impact computer security and critical infrastructures. Some of them were designed specifically to deal with security. Others were designed with different goals in mind, and such policies and laws have indirect effects (for good and ill) on security that need to be dealt with. Some of the legislation and policies in this category are discussed in more detail elsewhere in this book, including Chapter 3.

Security and cybersecurity legislation and policy is an area undergoing significant change in the United States following the September 2001 attacks and the development of the Department of Homeland Security. In this era of change it is even more important to think about secondary effects and unanticipated consequences of legislation and policies on computer security, and also how security-related legislation and policies could impact other areas of society. And with both the technology and use of computer networks frequently changing, legislation and policies will be lagging behind current activity for the foreseeable future.

The role of the private sector is more important as a result of these frequent changes in technology, since they typically lead the public sector in disseminating new technologies. The private sector also matters because a large percentage of computer networks and critical infrastructures are owned and/or operated by private companies. Regulating these companies is frowned on in the United States, in part because the information technology industry is credited with a great deal of the economic boom of the 1990s, and the parallel rise in American economic competitiveness. They are concerned that any attempts at regulating the industry will stifle its innovative output. While this is a point of view more sympathetic to the Republican Party, some Democrats, particularly those with many constituents in the industry, feel the same way. Others argue that the industry is not sufficiently responsive to the needs of consumers under current market rules. In California that line of reasoning contributed to the passage of a state statute (California Senate Bill #1386) requiring companies to notify their customers of a security breach that involved personal information of a citizen of California. Given the geographically disconnected nature of computer-enabled business, this could impact companies outside California as well.

While some advocates of government regulation may be seeking competitive advantage for particular companies or industries, others have consumer and other public interests in mind. Absent market-based incentives regulation is required if companies are to produce more secure products. A compromise would rely on various types of indirect means of encouraging private sector action. Some of these indirect means are discussed later in this chapter.

Product Liability

Software providers currently are not liable for flaws in the products they ship to customers. Given the history of the field, where new versions of products are often made available in beta form in part to have consumers find additional bugs, this isn't terribly surprising. In the minds of many, the lack of liability for software bugs and for faulty security is a major reason why computer security is not particularly strong. Increasing security, such as reducing the number of bugs in a software program, is considered an additional cost. If the company believes that the benefits of increasing security will outweigh the costs, then they will bear that expense. But most in the computer field either believe that the reward of producing more secure products does not outweigh the expense, or that the demand (via purchasing power, regulation, and/or lawsuits) from consumers for more secure products and processes is insufficient to take action.

Bruce Schneier, chief technology officer of Counterpane Security Inc., contends that "security is a process, not a product."[1] Part of that process is the context in which security decisions are made and the forces that constrain the choices made by producers and consumers of security products. Enforcing liability would dramatically change that context. The cost of not providing secure products would increase, hopefully to the point where companies would either produce those products or go out of business. Such liability could also extend to companies that control a large amount of data involving their customers or operate major computer networks, as well as to companies (perhaps both private and public) that provide critical infrastructures. Critical infrastructures (networks, both virtual and physical, that are essential to services such as transportation or power distribution) can create significant risks and losses if they fail. The greater risk may provide additional reason to enforce liability for computer security failures.

With the current balance of power and interest between the federal government and private industry, it's unclear how liability would be enforced. The industry exerted pressure on the Bush Administration to water down the recommendations in its National Strategy to Secure Cyberspace, and would likely resist efforts to revise product liability statutes. The Department of the Interior was forced twice within two years to disconnect from the Internet due to computer security issues, yet the government has not sought greater accountability from companies that provide its computer systems or computer security. Nor have there been similar actions against companies that exposed their customers (or their personal information) as a result of security problems. The balance of political power appears to favor those who support limited (or no) regulation. These individuals believe that because computers and computer security is a market good, the market will be able to find the appropriate solution.

A stronger argument for changing liability obligations suggests that computer networks function as a public good (a situation where government intervention is more accepted). These networks rely on participation by customers and employees to be effective and responsive; and they provide a service that provides benefits in part because they have lots of people participating in them (economists call this a network externality). Those parties that maintain the networks should feel some obligation to their customers, which in many cases include the general public. This line of reasoning is more persuasive for policies involving critical infrastructures, which often provide services (energy, transportation) regulated by governments. For example, the public good/public utility argument would support requiring minimum standards for server and router security, as those components

1. Schneier, Bruce. *Crypto-Gram Newsletter*, May 15, 2000. *http://www.counterpane.com/crypto-gram-0005.html#ComputerSecurityWillWeEverLearn*, accessed May 29, 2003.

service the public. Similarly, requiring computer products to be shipped with security settings on by default would help increase security and reduce the spread of computer viruses, supporting a public good. Considering the Internet as a kind of public utility is a shift from current and past perceptions of the Internet. It reflects the changing nature of both those who interact with computer networks and the continued spread of these networks. This is not a change in technology, but of rhetoric, perception, and policy.

If the government is unwilling or unable to mandate changes in product liability, there may be means to encourage companies to be more responsible for flaws in their software or their products. One scenario involves an attack that leads to significant losses for a company or companies. This would prompt the affected industry to police their own actions, or the government would insist on it. Hopefully it would not come to this, but the calls for greater security following the 2001 Code Red attack prompted little change, as evidenced by a repeat of the same calls when the Slammer worm hit in 2003.

Standards and standards bodies may be another means for enforcing liability, or at least encouraging greater industry responsibility for software and networks. This could be standards bodies of those respective industries, or of industries that use their products. For example, if the American Bar Association were to stipulate that any member of their organization could only use computer networks and software in their business that meet a specific set of criteria for security, then companies that do not meet those criteria will lose customers. If enough organizations that purchase software and utilize networks take such steps, Information Technology (IT) companies would need to follow suit to maintain their sales. The government could suggest or impose its own standards, which may be preferable if it is felt that industry cannot develop a set of standards that doesn't favor a particular company or companies.

Information Sharing and Analysis Centers (ISACs) could facilitate sharing of information as a way to increase network security. Created by Presidential Decision Directive 63 in the Clinton Administration (and amplified by Executive Order 13231 in the Bush Administration), participation in ISACs is voluntary, though encouraged. The centers report physical and cybersecurity threats to the National Infrastructure Protection Center, currently located in the Department of Homeland Security. But limited resources and persistent concerns about litigation or loss of competitive advantage from sharing information has contributed to the limited role of ISACs. Another business model or form of organization may be able to fill the role intended for the ISACs, but would still require greater participation from industry than the ISACs have enjoyed.

Some private sector companies are trying to be out in front on increasing security for the industry. In 2003, the Information Systems Security Association (ISSA) (*http://www.issa.org*) pitched a revival of what is called the Generally Accepted Information Security Principles (GAISP). The idea has been around since 1990, though under different names. It would be an information security version of the Generally Accepted Accounting Principles that U.S. corporations must follow when submitting financial reports. Different portions of GAISP would provide guidelines for top leadership, operations managers, and information security practitioners. ISSA plans to revisit and revise the document frequently. While chief executive officers (CEOs) are resisting additional regulations, some chief information officers (CIOs) are advocating regulation, or at least the threat of regulation, to prompt their companies into action. Self-regulation could be an effective way to postpone or prevent the imposition of government regulation.

Insurance is another indirect method of encouraging more private sector responsibility. While those favoring changes in liability law anticipate that insurance companies would encourage increased product security in light of the increased risk, such steps might occur without changes in legislation. If computer companies cannot be sued for faulty products and networks, those individuals and businesses that suffer losses as a result of security problems could sue the owners of the networks or the software that contributed to the losses. It is this type of indirect recovery that motivates the California law requiring consumer notification of security breaches. If sufficient cases alleging negligence due to inadequate or faulty security are filed and judgments rendered, insurance companies would take action to reduce their exposure and payout from these losses. If cases involving wrongful death (say if someone had died in the city where the 911 system was attacked by Slammer) were filed, that might provoke a faster response.

Insurance is already a factor for cybersecurity with some firms. Cyber-insurance policies have been developed for firms across all industries to cover things such as data destruction, business interruption, and various types of cyber attacks. Some insurance companies have cut hacking losses from general-liability policies, forcing companies to purchase additional "network risk" insurance. The cost in 2003 ranged from $5,000–$30,000 for every $1 million in coverage. It's unclear that this additional expense would be enough for some companies to change their practices. And like individuals with pre-existing conditions that have trouble obtaining health insurance, companies with troublesome security practices may find it difficult to obtain cyber insurance. Some companies have sued insurance providers for hacking-related claims—an additional expense that would prompt many to purchase the additional insurance. Companies have been slow to purchase these policies, with sales of less than $100 million in 2002. This is a relatively new form of insurance, and sales may jump once the industry is better acquainted with the special nature of underwriting cybersecurity risks. If cybersecurity insurance was required (much like automobile insurance is required to drive in some states), it would be an incentive for both the insurance and computer industries to better assess risks and losses from security breaches.

The absence of product liability may be an unintended consequence of the patchwork nature of software development. It is certainly a consequence of end user license agreements (also called shrink-wrap licenses) that exempt computer companies from liability. Consumers cannot negotiate the agreement; if they want to use the product, they must consent to the license terms. The absence of liability also allows for security risks to go undetected until it might be too late. Customers have come to expect software will be released with bugs. The complex nature of programming makes it difficult to thoroughly test software, much less see how it might function differently in a particular network environment.

The average consumer should not bear the burden of making sure that the products they use are secure. Most people who utilize information technology today do not fully understand how it works—a far cry from the early days of the industry or the technology. Expectations and experiences of the customers have changed, and the responsibilities of software and network providers should follow suit. One possible response is the development of smart failure systems, such as the recovery-oriented computing under development by researchers at Stanford and the University of California at Berkeley. Such systems recognize that errors (or attacks) are inevitable, and try to design systems to recover faster and restrict damage to the network when the attacks or failures happen. Another response is a more unified approach to security, integrating computer and physical security. In other words, the security people watching the entrance to a company's building would also monitor activity on their computer networks. This could lead to an increase of outsourcing computer security; a step Merrill Lynch took in early 2003. Assuming some combination of these possibilities and others, a more proactive approach to computer security seems likely. It's unclear who will pay for it and how it will be implemented.

An argument against increased regulation of computer companies is that they should not be unduly hindered from conducting their business and creating innovations. This is a common argument of economic conservatives and the Republican Party. The industry was a force behind the economic expansion of the 1990s in the United States. But the value of information technology to the economy will be compromised if security flaws continue to be exploited, resulting in loss of time, sales, and trust. The argument that regulation will hinder the progress of the industry was more persuasive when it was still young. As it now consists of many large companies and has oligopolistic tendencies, the industry should be sufficiently mature to adapt to changes in government regulation. If it isn't, then a change in the landscape may provide opportunities for other companies to enter, provide more secure products, and gain market share. Continued reluctance (or inaction) by firms to improve their security practices could lead to loss of market share and/or imposition of outside regulation.

Serious cyberattacks have yet to take place, and some of the people arguing that they are inevitable are construed as pessimists. However, computer security should not continue with an outdated fortress mentality or a belief that current levels of security are sufficient. In many ways computer security is like public health—the threats are constant, changing, and widespread. Prevention, detection, and response are three key parts of network security. Waiting for the next attack to take action is squandering the opportunity to minimize or perhaps prevent that attack from happening. Some in private industry have noticed the trends, as well as the turmoil in federal cybersecurity leadership and are acting before terms are dictated to them.

Government Actors

The federal government has taken steps to boost the security of its own computer systems and networks over the past several years. In addition to the 2003 National Strategy to Secure Cyberspace (discussed in more detail later in this chapter) several pieces of legislation were passed to better coordinate government cybersecurity activities. Relevant committees, of which there are several, are described in Box 1. While the number of different committees involved does demonstrate congressional interest in the issue, it also complicates the process of developing and passing legislation, since several different committees could claim jurisdiction.

The Office of Management and Budget is responsible for coordinating Executive Branch efforts in evaluating and improving government cybersecurity. The Executive Branch responsibilities for cybersecurity are currently located in the Information Analysis and Infrastructure Protection Directorate in the Department of Homeland Security. This Directorate now has the responsibilities of the FBI's National Infrastructure Protection Board (NIPB), the Commerce Department's Critical Infrastructure Assurance Office (CIAO), and several other agencies with responsibilities for cybersecurity and physical security. The removal of cybersecurity responsibilities from the president's office, along with the departure of two cybersecurity "czars" in short succession, added to the impression that the Bush Administration was not serious about cybersecurity. This perception was held by many who also believed that national policy should be more assertive in requiring action from the private sector.

Aside from the Department of Homeland Security, there are federal advisory bodies that contribute to cybersecurity policy. Advisory boards have also been helpful in developing policies for critical infrastructures. One such board is the President's Information Technology Advisory Committee (PITAC). These boards change from time to time, and from administration to administration. They serve as a means for people with expertise and/or significant interest in the issues to provide input to the federal government.

Legislation

The Computer Security Act of 1987 was the first legislation to address computer security in the federal government. However, some agencies took years to comply

Box 1: Subcommittees Involved with Cybersecurity in the U.S. Congress
(Full Committee in Parentheses)

Senate

- Subcommittee on Technology, Terrorism and Government Information (Judiciary)
  http://judiciary.senate.gov/subcommittees/technology.cfm

- Subcommittee on Homeland Security (Appropriations)
  http://appropriations.senate.gov/subcommittees/homeland/topics.cfm?code=homeland

- Subcommittee on Science, Technology and Space (Commerce, Science and Transportation)
  http://www.commerce.senate.gov/subcommittees/science.cfm

House of Representatives

- Subcommittee on Cybersecurity, Science, and Research and Development (Select Committee on Homeland Security)
  http://hsc.house.gov/content.cfm?id=18

- Subcommittee on Homeland Security (Appropriations)
  http://www.house.gov/appropriations/sub.htm

- Subcommittee on Research (Science)
  http://www.house.gov/science/committeeinfo/members/research/index.htm

- Subcommittee on Technology, Information Policy, Intergovernmental Relations and the Census (Government Reform)
  http://reform.house.gov/TIPRC/

---

with the act. Introduced in part to respond to the Reagan Administration's National Security Decision Directive that gave the National Security Agency (NSA) control over government computer systems containing "sensitive but unclassified" information, the act reasserted that the National Institute of Standards and Technology (NIST, then the National Bureau of Standards) would be the agency in charge of the security for unclassified, non-military computer systems in the federal government.

The NSA has tried since the passage of this legislation to regain control over security for governmental and non-governmental systems. Given the priority that agency places on total security (for good reason), it would be best for a civilian agency to have control in order to preserve public access to government systems and data and to ensure the robustness of non-military networks. National security interests highlight the tension between security and access that will be discussed later in this chapter.

The Computer Security Act established an advisory board, the Computer System Security and Privacy Board. The board was charged with examining issues affecting the security and privacy of sensitive but unclassified information in federal computer and telecommunications systems. The board's authority does not extend to private sector systems or federal systems that process classified information. The Act further required agencies to identify what computer systems contain sensitive information, make security plans to protect those systems, and develop training for those involved in the management, operation, and use of federal computer systems.

The Government Information Security Reform Act of 2000 (GISRA) consolidated many federal security policies. One of its provisions was a mandate for procedures to detect, report, and respond to computer security incidents. The law also required federal agencies to undergo both internal and external assessments of their security policies and practices. The legislation did not mandate specific agency practices or policies, nor did it mandate that best practices must be followed. The Office of Management and Budget (OMB) receives reports on these assessments, as well as plans developed by the agencies on how they would respond to weaknesses brought out by the assessments. These plans include milestones and note what types of resources they expect are required to fix the weaknesses. They are supplemented with quarterly updates. Each agency plan is grouped with other agency plans in order to give OMB, the agencies, and Congress a better understanding of the security policies and practices of the federal government.

Initial assessments indicated that the federal government still has several weaknesses in computer security. While the combined report to Congress was criticized for the relatively poor state of security it reported, it provided a baseline for comparison not previously available. Subsequent reports indicated that Federal IT security was improving, but is still not satisfactory. Performance in this area ranges widely across agencies, and as of this writing assessments have yet to be conducted in every federal system as required by law.

GISRA was replaced by the Federal Information Security Management Act of 2002 (FISMA). It permanently reauthorized the Government Information Security Reform Act and adds additional provisions. The assessments in the previous legislation were made annual, and the law requires agencies to follow the practices, standards, and tools developed by NIST (something that was optional in previous legislation). It also creates an Office of Information Security Programs to assist with this process.

It is unclear if the assessments required in the legislation, which includes an indication of the resources required for fixing the problems, will be met with the necessary funding. This legislation does make permanent an annual process of evaluation and planning meant to improve federal computer security, and encourages the development of best practices and policies for federal agencies. There are early indications that the plans and reports required by GISRA and FISMA are serving as both measuring tools and planning mechanisms that can help incorporate security concerns into management thinking and budget planning. As with other security legislation, full implementation is slow in coming, but it is coming. While this doesn't prove causation, it is important to note that federal government computers were relatively unaffected by the Slammer worm. Perhaps, as is hoped in the final draft of the National Strategy to Secure Cyberspace, the private sector will follow the lead of the federal government in developing assessment tools and utilizing best practices. But the industry appears to have the upper hand in this relationship and it may be more inclined to change only if economic forces prompt it.

The National Cyber Security Leadership Act was introduced in 2003. In substance it is very similar to both GISRA and FISMA, requiring assessments from the chief information office of each agency on the security of their computer systems. The systems assessed would include lower-level systems such as those targeted by the Slammer worm. These assessments would inform a standard setting process that would be handled by NIST to establish higher standards than those currently in place would. The legislation was introduced in part over concerns that agencies continued to receive failing grades in their security assessments. However, the assessments established by GISRA and FISMA had been required for only two years at the time this legislation was introduced. Raising the bar does not necessarily motivate agencies to improve faster.

The Cyber Security Information Act was introduced in part to carve out an exemption in the Freedom of Information Act (FOIA) for information concerning security vulnerabilities. It is a relatively broad exemption, covering any information that is designated a "cyber security data gathering." Such gatherings are not limited to cybersecurity information, and may include disclosures on critical infrastructures (which are not restricted to computer information under the proposed exemption). A different approach was taken by President George W. Bush when he signed an Executive Order allowing information that includes details about infrastructure to be classified confidential, secret, or top secret. Companies are reluctant to share vulnerability information because of concerns that it will be used by competitors. In light of the increase in terrorist activity, there is also concern that terrorists would use that information for their advantage. By providing an exemption to FOIA (or classifying this information) industry may be more willing to provide information, which could be useful in assessing computer security and may point out problems not otherwise easily determined. However, it reduces the ability of the public and public interest groups to monitor the actions of industry.

Exemptions to FOIA, or increased classification of information, should not be taken lightly. FOIA is an effective means for the public to obtain information from the government, and any exemption weakens that avenue of public oversight. Such an exemption would prevent security researchers, even those not affiliated with the private sector, from using the information in their research. There are already Information Sharing and Analysis Centers (ISACs), created by Presidential Decision Directive 63, that share critical information in a confidential manner, though private companies are still reluctant to cooperate for fear of losing the advantage from any proprietary knowledge shared in the ISACs. The vague definitions of what cybersecurity data is under the exemption would allow companies to claim exemptions for data not directly related to security threats or vulnerabilities, thereby providing too much protection from public scrutiny. As the pressure for a FOIA exemption or increased classification is significant, the definitions and processes involved in designating such material are important in determining if security interests are really being served by granting an exemption.

Other Legislation with Security Implications

What follows are two examples of legislation with secondary, or even unintended consequences, on computer security research. This reflects the interconnectedness and complexity of both computer systems and the policy process. There is of course legislation that affects cybersecurity, but through indirect means, such as the USA PATRIOT Act, which increases penalties for crimes such as hacking, and expands the information technology tools and means of surveillance available to law enforcement. Other chapters throughout this book mention legislation that can impact the ability of consumers to obtain secure products, have their personal information protected online, or allow for the appropriate amount of trained workers to support secure computing and critical infrastructures.

Digital Millennium Copyright Act (DMCA)

Digital Millennium Copyright Act (DMCA) legislation was passed in 1998, and is given a more thorough examination in Chapter 7. The DMCA was the first major effort to stop the circumvention of anti-piracy tools. There are two kinds of prohibitions the legislation has incorporated into the Copyright Act: acts of circumvention and acts of distributing tools and techniques that can be used for circumvention. These provisions, particularly the one concerning distribution of tools and techniques, create a tension between content providers and the security research community. "White hat" hackers or "red teams" often attack systems to seek out vulnerabilities and demonstrate flaws in systems, and such an exercise prompted the initial court action to disconnect the Interior Department from the Internet (Native American tribes were concerned that the Department was not effectively managing their assets).

This applies to anti-piracy systems as well. Given the difficulty of fully and rigorously testing computer systems prior to deployment (and the lack of incentive for businesses to do so), sometimes these tactics are the only way to find particular problems in software and hardware. However, at least in the United States the interests of copyright holders to protect their works appears to take priority over research on the vulnerabilities of anti-piracy means. The anti-circumvention provisions of the DMCA have been aggressively enforced in a few cases discussed in more detail in Chapter 7.

Security research is not exclusive to the academic community. Many businesses are engaged in such research, and companies may be tempted to utilize DMCA to stifle competition. Former head of the White House Office of Computer Security Richard Clarke is on record saying "a lot of people didn't realize that [DMCA] would have this potential chilling effect on vulnerability research."[1] The DMCA could be utilized to have the government limit competition in the name of protecting intellectual property and ensuring security. Enforcement and interpretation of this statute will continue to have an influence on security research in both the private and public sectors for the foreseeable future. It is one of several trends leading toward continued support of content providers that will restrict fair use and other activities that could also hinder security research.

Uniform Computer Information Transactions Act (UCITA)

The Uniform Computer Information Transactions Act (UCITA) is a proposed contract law introduced in 1999 that would set new rules for licensing software and several other forms of digital information, and is discussed in greater detail in Chapter 7. The restrictions on the use of intellectual property concern a restriction of fair use (among other concerns); the impact on security concerns is not as obvious. It comes from the continued shift in power toward software providers in the negotiations over licenses. Vendors can, even under current law, insert terms that would obligate licensees to permit the vendor to remotely install periodic updates to their software. They could do this even if the update disabled other software in the computer system. This is a problem found in many software patches as well. While the patch does fix the security risk, it can disable other software. It's done unintentionally, because software is typically not tested within integrated systems. Such testing is an incredibly difficult thing to do, given the

---

[1].   Bray, Hiawatha. "Cyber Chief Speaks on Data Network Security," *Boston Globe*, October 17, 2002, C2.

complexity that results from several different software programs interacting with each other. The perspective of many vendors is *caveat emptor*—let the buyer beware. Other practices, such as shipping components of wireless networks with the security settings off, reflect the same perspective.

National Strategy to Secure Cyberspace

Issued in draft form by the President's Critical Infrastructure Protection Board in November 2002, the National Strategy to Secure Cyberspace (the Strategy) is the major document outlining federal government efforts to raise awareness about cybersecurity and encourage action at all levels of computer operation. Such a plan was in the works since shortly after the September 2001 terrorist attacks. The draft plan differs significantly from the final document issued in February 2003. A shift in cybersecurity authority from the Office of the President to the Department of Homeland Security and the departure of Richard Clarke, the major architect of the draft strategy, also contributed to a perception that the final plan lacked teeth as well as someone or some organization with the political capital to implement the plan.

The draft report was organized around recommendations aimed at five different levels of computer users, including individual home users and small businesses, as well as recommendations for national priorities and global issues. It encouraged the use of market forces to motivate information technology companies to produce more secure systems and otherwise make consumers more aware of existing security features. Such measures are not found within the final plan, which is more focused on what the federal government needs to do to improve its own cybersecurity and to suggest (not encourage) others to follow suit. The number of recommendations is down, especially for industry. The final report outlines five national priorities, encouraging the use of public-private partnerships to assist as needed. The priorities are:

- A National Cyberspace Security Response System

- A National Cyberspace Security Threat and Vulnerability Reduction Program

- A National Cyberspace Security Awareness and Training Program

- Securing Government's Cyberspace

- National Security and International Cyberspace Security Cooperation

While the levels found within the draft plan were not abandoned (they are now under the first priority); the general tone of the final version is hopeful that industry will follow the example of the government. While the government may not feel it has the authority or the expertise to dictate what private entities should do, it could have argued more explicitly for action to improve cybersecurity on behalf of the public trust. The draft plan is certainly more ambitious and assertive in that regard. Hopefully the National Strategy will be periodically revisited, and evolve with changes in technology and practice. That is good practice in any area of policy where the technology and the use of that technology change often.

With the information technology industry leading the way in many areas of the field, it would be difficult for the government to effect dramatic change through leading by example, which is the major thrust of many of the recommendations. Security is not a field where the public sector is out in front. Arguably nobody is really out in front on cybersecurity due to the lack of market incentives to provide more secure products and ensure secure networks. But most of the activity in computers, including security, does not take place in the government.

Flaws

There are two major problems in both the draft and final versions of the strategy. Both are couched in terms of recommendations without detailed action plans or budgets. The final strategy does recommend the development of a research and development agenda, but fails to suggest recommended dollar amounts or preferences about who should perform this research. Since the Bush Administration made performance-based measurement a key part of its budgetary process, the absence of more concrete action plans and performance goals is curious.

The final plan provides a good start for improving the national cybersecurity infrastructure. A major issue in cybersecurity debates is the lack of information (especially public information) on cyberattacks and other security risks. The recommendations for developing a response system, raising awareness, and improving monitoring capabilities (through the Cyber Warning and Information Network, a complement to the Global Early Warning Information System being developed by the National Communication System) recognize the paucity of information on cyberattacks and related network difficulties. While critical infrastructure policies of the late 1990s encouraged the development of Information Security Assurance Centers, greater use of the ISACs or a more widespread cybersecurity infrastructure could certainly be useful in helping government and industry better monitor, detect, and respond to attacks. The nation's proposed research and development strategies are also encouraging in their cybersecurity efforts, though some were concerned that policy pronouncements and legislation were not being supported by sufficient budget and other agency action, both within the Department of Homeland Security and in other agencies.

There is a particular tone to both the draft and final versions of the Strategy that could do as much harm as good. Both are written primarily from the standpoint of reducing the threat of cyberterrorism. This is a bit of a red flag to some in the security field because the nature of the cyberterrorist threat is largely unproven. The potential is certainly great, and nothing to be dismissed out of hand. But the absence of documented cyberterrorist attacks makes some question the likelihood of an attack. Terrorists at this point are probably more likely to rely on cyberspace in support of their operations than to try and take out major computer systems. For all of the talk about cyberwarfare around the time of the 2003 conflict with Iraq (and the apparent support of retaliatory attack in the National Strategy), a similar lack of known, successful, and significant activity supports critics' arguments that the risks are overblown. Lost in the rhetoric in this debate (Richard Clarke being a noted proponent of preparing for an electronic Pearl Harbor, and Declan McCullagh one of many who suggest such a stance is an overreaction) is a serious discussion of risk assessment and analysis, and the need for such data on attacks in order to conduct such an analysis.

Defense or Service?

The shift of cybersecurity boards and offices into the Department of Homeland Security, and a concentration of cybersecurity activity in that department portray the role of cybersecurity as protecting the homeland instead of ensuring a needed service that supports all facets of life. To cast the function of cybersecurity entirely with a defense-oriented agency is problematic. As mentioned earlier in this chapter, a national security perspective encourages a strong and total defense against outside threats. Given the stakes of national security interests, anything less than total security results in failure and loss. While total security for a defense agency is perhaps a necessary thing, the robust, complex, dynamic nature of computer networks makes such security impossible to guarantee, and counterproductive to other interests in a public system. The Department of Homeland Security should remember the porous nature of computer networks and how that porous quality is often what makes such networks so valuable. The Department of Commerce, while not as focused on security concerns, may be better prepared to recognize and embrace the imperfectly defendable nature of the Internet.

It would be less alarmist and a better persuasive strategy to recast the need for increased cybersecurity in terms of reliability. Bill Gates, in a memo outlining Microsoft's Trustworthy Computing initiative, makes an important parallel to public utilities when discussing cybersecurity. He believes that computers should be as reliable as water and electricity systems. Improving cybersecurity can reduce network slowdowns and failures, and bring those networks to the higher reliability standards of public utilities (which also rely on both humans and technology for that reliability). Consumers don't take kindly to interruptions in water service, and they shouldn't have to tolerate the degree of computer service breakdowns they currently do. Efforts like recovery oriented computing could help increase reliability, or at least trust in the system.

Such an approach to cybersecurity may involve a greater role for other federal agencies like the Commerce Department. It would also cast cybersecurity in terms of prevention and detection as well as response. For example, the Federal Trade Commission supported the Common Sense Guide for Home and Individual Users released by the Internet Security Alliance. The report was a part of a series of guides geared toward different classes of computer users. They promote cybersecurity through education and recommendations. Their main thrust is increasing consumer savvy about the Internet, prompting individuals and businesses to ask the questions they don't think about, or assume that their IT department handles. By casting the goals of cybersecurity more broadly than defending against terrorists and other enemies to include ensuring the health and welfare of computer systems, the government could develop a broader, more inclusive approach to cybersecurity that preserves the robust nature of our computer networks.

The strategy could also focus more attention on the human components of computer security, and the common tactic of hackers to gain access through manipulating the people involved in the computer networks they want to access. Such social engineering is an important part of security, and can be quite powerful. Noted hacker Kevin Mitnick's ability to utilize social engineering concerned the judge in his case so much that he was placed in solitary confinement. Education and awareness of social engineering is not addressed in great detail by national cybersecurity documents outside of exhortations to increase awareness of security issues.

Implementation

The Strategy is a good step toward improving the nation's cybersecurity. A critical challenge is implementing the Strategy. The government has responded positively to the Strategy and the security-related legislation that preceded it. But the private sector must be engaged with this plan in order for it to become part of national computer practices. They resisted the mandates in the draft Strategy, concerned that regulation would be hinder industry. While the government could encourage increased standards through its purchase requirements, it is unclear if the government has enough market share (or market influence) to ensure that increased standards are made available to other consumers. The information technology industry, much like the computer networks that contributed to its success, arose in an environment of limited or non-existent regulation. They feel that any regulation would hamper innovation, but an industry as mature as information technology is could handle a bit of regulation. Other critical infrastructures such as banking have been regulated. Such regulations have prompted needed investments and some see such regulation as a means to stimulate additional security measures. But the current Strategy is little more than asking the private sector very nicely to improve their security. Past actions indicate such moral suasion will likely prove ineffective.

The government will need credible, forceful individuals to help make the Strategy a reality. The placement of cybersecurity responsibility further from the president will make it harder to get things done because of a reduction in access and influence. However, the relative power of an agency does not always coincide with the threats it is obligated to face. The government should continually revisit the Strategy to respond to changes in the cybersecurity landscape. It must also be careful not to dismiss or downplay other important governmental interests in the pursuit of cybersecurity. The Information Analysis and Infrastructure Protection Directorate of the Department of Homeland Security will hopefully lead national efforts, but it should not be the only government voice on cybersecurity. Input from many different parts of the government (and the country as a whole) would help negotiate the tensions between appropriate precautions and actions to ensure cybersecurity or encouraging a pessimistic mentality that diminishes computer interconnections and their associated benefits.

The government can be successful in improving cybersecurity. In 2003 a security company in Atlanta discovered a major flaw in an email program that powers more than half of the world's email. The company notified the Department of Homeland Security, which in turn talked with many vendors that bundle the email program with their products. The two organizations worked together to help fix national security, military, and other federal computer systems. This was all done under non-disclosure agreements that were not violated. Such agreements were no doubt necessary due to industry's reluctance to share vulnerability information prior to the development of a patch. The true success of the patch for this flaw still comes down to individual companies and users incorporating the patch into their systems. But industry cooperation with the government is a distinct departure from the general mistrust of the government from the computer and security industries. That distrust is a big challenge to the successful implementation of the Strategy.

Research and Development

A good portion of the debate about cybersecurity and policies to improve it concerns what actions need to take place and what products need to be developed. Not as much attention is given to what we know about computer security and information assurance, what research is being done in the field, and how many people are employed in both research and application of cybersecurity. This lack of knowledge prevents much of the cybersecurity debate from moving from possibilities toward actions, and arguably prevents any comprehensive cybersecurity strategy from being developed and implemented successfully. Most security policies are reactive rather than proactive, and increases in computer security knowledge typically happen after attacks or circumventions of security measures, not before. The average user may not see the lack of knowledge as a problem (or believe such a lack exists), but those that rely on cybersystems for their livelihood or are trying to define the risks involved really need more information. The Strategy recognizes a need for more research and development in cybersecurity, and there are two areas in which more people and funding is needed: personnel and research.

The personnel needs are two-fold. More people are needed in universities and industry for researching computer security (approximately seven PhDs are granted each year in computer security–related areas), and people trained in cybersecurity are needed in government and industry. The people for the second category need not be trained at the Ph.D. level, but bachelor's and master's degree programs in cybersecurity are relatively few. The Scholarship for Service program is one attempt to increase the number of computer security professionals. In 13 schools as of 2003, the program provides scholarships for students in graduate and undergraduate programs studying information assurance in exchange for one to two years service in the federal Cyber Corps following graduation. Part of the program includes summer internships with government agencies. The first students graduated in 2003. Graduates can be placed with federal, state, and local government agencies. These people could assist with updating national computer security standards, ensuring the viability of a local city's electrical grid, or any number of computer security functions in the public sector. Industry would do well to utilize the talents of these people once their government service is concluded, because IT personnel in industry are often overextended in their own responsibilities, and many do not have training in cybersecurity. Similar programs or fellowships could be supported by industry, but it's unclear if the incentives are there to commit the necessary resources.

Similar training in cybersecurity takes place through exercises like the Cyber Defense Exercise organized by the military. Similar to the capture-the-flag competition at the annual Def Con hacker convention, the Cyber Defense Exercise has teams defending and attacking isolated networks as a real-world activity in cyber-attack and cyber-defense. It helps demonstrate the asymmetric nature of cybersecurity (to succeed, those on the attack need find only one flaw, while defenders must protect against anything) and provides experience hard to teach in a course. Similar "red team"-style programs should be encouraged.

The second major area for research and development is increasing knowledge of cybersecurity. National spending on cybersecurity research was around

$60 million in Fiscal Year (FY) 2003. By comparison, $59 billion in new information technology spending was in the Bush Administration's FY 2004 budget request, and the research and development money targeted toward bioterror-related research is in the billions as well. The Cybersecurity Research and Development Act was a first step in ramping up the cybersecurity research and development agenda. Signed in 2002, it provides over $900 million in research grants, fellowships, and development of computer security research centers. The National Science Foundation and the National Institutes of Standards and Technology will administer the research grants. While the legislation was passed, there were complaints that agencies were slow to implement the legislation.

There are many areas deserving attention in the cybersecurity research and development agenda. The Strategy identifies some existing priorities, including intrusion detection, Internet infrastructure security, application security, denial of service (DoS) attacks, and communications security. The Strategy does not address research on the human and organizational elements of cybersecurity (such as social engineering), but does charge the Department of Homeland Security with coordinating research between universities, industry and government. The Institute for Information Infrastructure Protection (I3P), a non-profit consortium dedicated to cybersecurity and critical infrastructure protection, identified some areas for new or additional research: enterprise security management; trust models for distributed parties; wireless security; metrics and models; secure system and network response and recovery; and law, social, and economic issues. The Institute also noted the needs and challenges in improving the flow of ideas and technologies between researchers, developers, system integrators, and end users. The Institute's approach to cybersecurity is a stronger acknowledgment than the government's that cyberspace is a series of systems that interact with each other and with systems outside of cyberspace. It certainly is more explicit in its recognition of the human element in security. Again, security is not a product (or a specific technology), but a process that depends on people and technology.

The Cybersecurity Research and Development Act should be the first step in a long-term research and development strategy. Training people takes time, and even with significantly more resources, the immediate impacts of such research and development will be small. To be truly effective, coordination between government, industry and universities must be encouraged in order to best leverage the efforts in one sector with the activities in the others. The knowledge developed through these training and research efforts will need to be shared. The public-private partnerships encouraged by national policy are helpful in this regard. Given the tensions between government and industry over the sharing of cybersecurity information, this may be a significant challenge.

Such research can go a long way in improving computer security and in increasing public recognition of the multitude of secondary and unintended effects of computer networks. The knowledge this research provides can help develop new security models to replace the outdated and inadequate Maginot Line perimeter defenses currently in place. It can redefine cybersecurity in a way that fully embraces its technical and non-technical components, and provide the expertise and tools needed to shift from a passive, reactive defense to an active one. As next-generation networks are being developed and citizens' lives become increasingly dependent on computer networks (networks they expect to be secure), research and development is an essential complement to national strategies and tactics to support cybersecurity.

Critical Infrastructures

Computer networks like the Internet are a critical infrastructure. They provide the backbone and supporting mechanisms that allow information to flow, electronic transactions to happen, and computers to "talk" to each other. Infrastructures are networks, both virtual and physical, that are essential to services like making the trains run or keeping the lights on. People probably don't think too much about these infrastructures, but they rely on them everyday. They include: agriculture and food, water, public health, emergency services (police, fire, rescue), defense, telecommunications, energy, transportation, banking and finance, chemicals and hazardous materials, and postal and shipping networks. Computer networks are part of an infrastructure and also support infrastructures. Many critical infrastructures, like transportation and power networks, rely on computers to function effectively.

Each of these critical infrastructures is vulnerable to attack. These systems are vulnerable to computer security problems as well as attacks that target their physical structure. The Slammer worm affected two different critical infrastructures. Several Bank of America ATMs were disabled, and the 911 system in Bellevue, Washington had to resort to pencil and paper. The resulting slowdown of service put lives at risk. Critical infrastructures are also directly targeted for attack. In 1994 a computer in Arizona that runs one of the state's dams was hacked. The hacker was in critical areas of the dam's control system, but did not have any control over the floodgates. Approximately 70 percent of the nation's power plants reported being hacked in 2002. In each case, some network computer connected to the Internet was breached, and the critical infrastructures were then accessed via that computer.

It's important to emphasize the vulnerabilities of these systems, while appropriately assessing the difficulty in staging an attack on them that would have significant impact. If 70 percent of the nation's power plants are being hacked without considerable numbers of blackouts or other disruptions to power generation and distribution, then causing significant damage to these systems is not something the average hacker can do successfully. It takes more than a few keystrokes. But the difficulty of these kinds of attacks and the relative obscurity of many of these systems does not mean they can be dismissed as unworthy of attention.

For example, since many power Supervisory Control and Data Acquisition Systems (SCADAs) are connected to power company LANs, a successful intrusion on the SCADA via such a LAN could disrupt power for a town or region. If a virus spread through the financial services network and disabled major stock and commodities exchanges, even for a few hours, it would affect the prices of goods and the flow of money that is a characteristic of today's global economy. If the attackers take the time and effort to plan a systemic attack at several points (much more difficult than a single intrusion), or combine a cyber attack with a physical one, the resulting impact could be much wider. For other critical infrastructures powered by electricity that is not independently generated (some water treatment systems, for example), problems with the power grid also mean trouble distributing other vital goods and services.

The interlinked nature of many of these systems can be exploited via the Internet. In light of the geographically diffuse and remote nature of many of these systems, infrastructures are often controlled by digital control systems. Systems like SCADAs or Distributed Control Systems (DCSs) perform functions as simple as monitoring portions of a network and reporting back to a central control, to throwing switches on railroad tracks, to controlling the floodgates of a dam. The communications with these systems was over a telephone line, and is now done electronically in many cases. Also, the operating systems of these networks are not as specialized or customized as they were in the past. Standardization reduces expenses and makes it easier for someone to operate, for good or ill, within a number of different control systems. It is likely that the standardization and interconnectedness of these systems will continue, though these systems are not likely to be updated with the frequency that someone might upgrade their laptop, given the expense involved. With a relatively minor human presence, it is important that these systems be secure from both electronic intrusion (including wireless intrusions, or wardriving) and physical attack. They are both threats to the effective functioning of critical infrastructures. While the threat of cyberintrusions has increased, many experts believe it is still easier to bomb a control system than to hack it. That does not dismiss the threat of electronic attacks, but hopefully places it in a proper perspective.

History

Security risks to critical infrastructures were a government concern before the September 2001 attacks. The National Communications System (NCS) was established in 1963 following the Cuban Missile Crisis to ensure reliability and availability of national security and emergency preparedness communications. The "hotline" between the United States and the Soviet Union was a product of this commission. Subsequent administrations saw fit to expand the capabilities of the NCS. The private sector was brought into telecommunications infrastructure protection in the 1980s with the formation of the National Security Telecommunications Advisory Committee, an advisory board of telecommunications CEOs. They recommended the formation of the National Coordinating Center (NCC) in 1984. The NCC is a government-industry information sharing mechanism concerning telecommunications. It was designated as the ISAC for the telecommunications industry in 2000.

The release of the Morris Internet worm in 1988 prompted the first federal efforts concerning the infrastructure of computer networks. The CERT Coordination Center was established at Carnegie Mellon University to coordinate communications among experts and to help prevent security incidents. Other CERTs have been established around the world.

Perhaps in response to the 1995 bombing in Oklahoma City and the pending Y2K situation, the President's Commission on Critical Infrastructure Protection (PCCIP) was established. It was the first comprehensive effort to assess and address vulnerabilities in national critical infrastructures. In 1997 the PCCIP issued a report outlining a strategy for evaluating and protecting critical infrastructures from both physical and cyber threats. Their report emphasized the strong dependence on computers and computer networks in each of these systems. Responding to this report, in 1998 President Clinton signed Presidential Decision Directive 63, which mandated that agencies secure their own critical infrastructures and take the lead in private sector efforts. It also created various positions and boards to help coordinate national policy making in the area of critical infrastructure protection.

Response to the Directive was slow, in part due to its vaguely defined goals and standards. The CIAO in the Department of Commerce provided oversight to the effort, and the National Institutes of Standards and Technology eventually developed a team to provide assistance and support to agencies with scarce resources. A plan for critical infrastructure protection, the National Plan for Information Systems Protection Version 1.0, was released in 2000. Like many reports in the area of security, it encouraged private-public partnerships. Follow-up versions were intended to address the responsibilities and roles of state and local governments, but were superseded by a change in direction following the September 2001 attacks.

Private industry responded to these efforts by establishing sector-specific Information Sharing and Analysis Centers (ISACs) to assist in sharing information between the private sector and the government. The private sector also established the Partnership for Critical Infrastructure Security (PCIS), a public-private partnership represented by each of the critical infrastructure industries that is a forum for issues relating to infrastructure security. The private sector also had influence on the final version of the Strategy (as discussed earlier in the chapter).

The Bush Administration continued such efforts and incorporated critical infrastructure recommendations into the Strategy. President Bush issued Executive

Order 13231 to call for continued action and efforts along the lines of Clinton's Presidential Decision Directive 63. The President's Critical Infrastructure Protection Board was created as part of this Executive Order. Much of the responsibility for critical infrastructure security is now with the Department of Homeland Security, though many federal agencies have some responsibility to safeguard facilities and/or critical infrastructures. State and local governments are also involved in protecting and operating critical infrastructures, and in many cases will know more than the federal government about the facilities and services that need to be secured. These governments are also struggling to improve their first responder capabilities to address terrorist and other threats, and have financial constraints of their own. It is likely that efforts to better secure critical infrastructures will struggle with limited resources and cross-jurisdictional entanglements.

Tools

Policy tools that can be used in increasing security of critical infrastructures are the same as those about cybersecurity discussed elsewhere in this chapter. A key difference is the rationales available to justify using these tools. These infrastructures provide needed goods and/or services to the public, and their failure could have dramatic impacts on public health, safety, and the economy. Government intervention is more understandable when it applies to infrastructures with a clear public interest at stake.

ISACs can be an effective tool to help in detecting and responding to cybersecurity threats. ISACs have also been established in several infrastructures, including energy trading, finance and banking, food, pipelines, surface transportation, telecommunications, and electric power. The telecommunication sector was in front on cybersecurity in part because it had already established groups to assess and strengthen its networks. The key thing to remember is that many computer networks and critical infrastructures are managed, if not owned, by the private sector. Public-private partnerships like ISACs are a way to ensure communications between government and industry in times of crisis. However, without mutual trust and sharing of information such partnerships will not be effective, since parties will be less willing to participate in them.

One major question raised by security concerns about critical infrastructure is the need for these systems to be connected to the Internet. While there may be a need for the computers in these infrastructures to be networked, the advantage of being connected to the Internet, an inherently insecure system, does not outweigh the risks of exposure to access by a determined hacker, or even someone with the proper automated software and time on their hands. The military has many of its networks isolated from the larger Internet to minimize security risks. The vital nature of these infrastructures would justify making sure that such systems were similarly isolated. That may be easier for those under government control or regulation (energy utilities, the banking system), than those predominantly in private hands (shipping and transportation networks). The private companies handling critical infrastructures could take some lessons from the phone companies, who take pains to secure their facilities from intrusion. It should be noted that such measures will not stop attacks caused by insiders determined to harm the system, or by employees who inadvertently expose systems by connecting computers to the Internet and then reconnecting those computers to these systems without rechecking them for viruses or other security vulnerabilities.

The Future

The actions taken to date were effective first steps in increasing the security of critical infrastructures, in particular the computer components of these infrastructures. But these were first steps. Like the Strategy, the infrastructure protection plans were high on awareness raising and low on specific actions. These strategies were focused on government activities, leaving the private sector to act as a good citizen with little direction or regulation. Raising awareness is important, and the government is typically more effective in getting its own house in order than in encouraging the private sector to do the same. But education must be followed by action and accompanied by leadership.

The positions in the Department of Homeland Security related to cybersecurity and critical infrastructures were slow in initially being filled. They are also located in the middle levels of the Department, far from the Secretary. Without strong leaders to champion their cause, these parts of the Department can lose out in the budget process. Even if the Department of Homeland Security is fully funded and staffed with respect to critical infrastructures, several other departments and agencies must also be so staffed, funded, and trained. Industry also has responsibilities to consider, as do universities to a lesser extent (a point mentioned in the Strategy). Coordinating mechanisms like the ISACs could be helpful in making sure information flows in all the appropriate directions.

With critical infrastructures, the cybersecurity issues are not a subset of those for the Internet, but a special case. Like military networks, there are reasons for these systems to be connected with the Internet, but the risks (and the potential losses from an attack) are greater than with other systems. Like military networks (or even regular commercial networks), their interfaces with the "outside" are points of intrusion and access. And like military networks, infrastructure networks would be better off to minimize outside access as much as possible. Military networks are relatively isolated, and critical infrastructure networks were too, though that is changing

Unlike military networks, utilities and other private companies responsible for critical infrastructures are in a new era of deregulation, consolidation, and outsourcing. They are under pressure to cut costs, and Internet and wireless connections are ways to reduce costs and automate systems. The consolidation of firms and the increase of electronic control has provided fewer targets for attack, but also created more points of entry. Cutting costs is admirable, but these private companies have responsibilities to the public beyond providing the best service at the lowest cost. The public interests at stake in these systems also make it easier to advance more restrictive policies because the tension between the need for access and the need for security is not as strong. Unlike military networks, however, these systems are mainly privately owned.

Government actions by themselves will be insufficient to better protect critical infrastructures. Public-private partnerships in this area (like the PCIS) are progressing further than other cybersecurity efforts, and may serve as a model for other areas of security. Ultimately, the security of critical infrastructures does deserve attention, and could use improvement. But like many of the reassessments of security in the wake of the September 2001 attacks, proper risk and resource assessment is critical to ensure that things are protected in a manner befitting the risks of a successful attack or disruption. Being overly sensitive to security threats invites panic, which is counterproductive.

Analysis

Various issues and tensions involving cybersecurity and critical infrastructures have been raised or implied in this cursory examination of the background, relevant legislation, and policy tools involved. Many of these issues result from conflicting interests. Others are the result of the adversarial relationship between the private and public sectors in the United States, at least where information technology is concerned. Information technology raises a host of issues and challenges, and security intersects with several of them. This section will engage many of the issues touched on in this chapter and elsewhere in the book, with an eye to how the policy process is challenged or can assist in addressing these issues and resolving these tensions.

One theme that will hopefully be clear in this section is that much like how the interactions of several computers (or several pieces of software) within a system can generate unanticipated benefits and risks, so too can the intersection of several different policies. The diffuse nature of policy making in the U.S. government, with many different entities involved, can make it difficult to assess the impact of one policy on a different area, either before or after implementation. Not only can there be unintended consequences of a specific policy, but the effects of one policy can conflict with those of another.

Consider anti-piracy measures. The DMCA is intended to combat digital piracy of intellectual property. While those concerned with piracy have raised their own objections to the bill, there have been secondary impacts of the legislation. Dissemination of research on circumventing anti-piracy tools is illegal under the act, and has chilled such research. The hardware-based anti-piracy solutions encouraged by the legislation are not as flexible as software-based solutions. It would be harder to replace outmoded hardware than software, and this makes hardware-based protection systems less secure. There's enough of a challenge in getting computer users to install software patches. To force them to purchase new hardware every time a device is breached is unrealistic. Without significant coercive measures, consumers will most likely retain the old devices and place their systems at greater risk.

Similarly, policies that are intended to increase computer security have the potential to infringe on privacy considerations. Increased traffic analysis of computer networks would make the Internet much less anonymous than it has been. To what extent will privacy concerns be addressed in the development and implementation of monitoring and detection methods to increase computer security? What guarantees will users have that such information would be kept confidential? The affect on other areas of policy flows both from computer security policies and to computer security policies.

There are a number of interests to be kept in some kind of balance where computer security is concerned. The remainder of this section will discuss many of these balancing acts as well as the potential for policy to assist or hinder the resolution of these tensions.

The Best Is the Enemy of the Good

As mentioned earlier, computer security has a military or national security history. In those arenas the risks of failure were large enough to warrant significant expenditures and many levels of protection. The goal was total security, and while there are always holes in a plan, military and national security interests had the resources and the motivation to patch them and get very close to total security. They can build their high walls around their bases and systems, isolate them from the Internet, and otherwise try to keep everything out.

Other computer systems don't have that capability or wouldn't benefit from such a change. This is an issue of both resources and intent. Many users of computer networks and the Internet rely on the ability to access systems and interact with them. Many benefits of a network come from the distributed interconnectedness of various computers and their users. To disconnect these systems from the Internet would be crippling, and to have multiple levels of security to access their website would dissuade some customers and intended users from visiting computer networks.

This doesn't mean that computer systems can't be secure. It does mean that they can't be completely secure. Even with that in mind, many computer systems have inadequate security. Two major reasons why computer security is not a greater concern among consumers are that many attacks are not publicly known, and the effects of many attacks are often inconveniences rather than things that cause harm or loss.

Unfortunately, not all attacks are simple inconveniences, and as computer systems grow in size and extent of interconnectedness, the possibility a security breach that will result in financial ruin, identity theft, or loss of life remains. Steps can be taken to mitigate these risks and respond to failures, but they can never be eliminated. Policies and attitudes need to reflect this fact. Total security can be a goal, but it can never be fully realized. Systems will fail, so how they respond to failure should be part of their design and development. The Internet can route traffic around parts that suffer failure and/or congestion. Can a company's network do the same? Computer security (and security in general) should be developed with failure in mind. If a system fails, can it be prevented from affecting other systems? Can it fail in a smart way or be better designed for recovery, minimizing damage or other impacts? The interconnected nature of computer systems makes answering these questions essential, yet more complicated.

More Information Is Needed

We don't know enough about computer security to make effective policies. We don't know enough about what could realistically happen and we don't know enough about what has happened. That doesn't mean we shouldn't develop policies because the need for security is independent of our understanding of it. But the lack of knowledge and certainty makes it difficult to have a policy debate that doesn't result in the two sides trading arguments that are little more than guesses about what could or couldn't happen. We have worst case scenarios and possibilities, but little data on past attacks and other security tests to assign realistic probabilities of risk. As a result, security models have evolved little over the past several decades, and predictions of dire consequences are often met with comparisons to Chicken Little: The sky may not have fallen, but it could.

The Cybersecurity Research and Development Act is a good start in improving the knowledge base of computer security. Coupled with the priorities of the Strategy that encourage development of monitoring and detection capability, the United States is preparing a foundation of computer security knowledge and understanding that will help advance the debate and craft more responsive and effective policies to mitigate computer security risks. Such a foundation must be crafted with the private sector to make sure the knowledge is well distributed and that the private sector in information technology comes to trust the government more than it has in the past.

It may never be possible to determine all threats to computer security and/or always devise appropriate responses to those threats. The technology changes quickly enough that security measures will always be changing. There is also the uncertainty inherent in combining several hardware and software components together in the field. Rigorous testing can provide some insight into potential bugs and other risks, but the complex interactions found in computer networks will provide their own unique surprises. Computer security processes should keep this uncertainty in mind, and focus as much on strategy for computer security as on tactics. It should never be assumed that security risks will only come from certain places or only through certain parts of a computer network. This is a troubling aspect of the Strategy. By casting computer security predominantly in terms of homeland security, it discounts other sources of computer security threats. Doctrine and threat scenarios can be developed and practiced, but it is just as important to be able to respond to the unexpected threat as it is those you can expect.

Merchants and Guardians

Scott Pace, in a monograph of the same title,[1] describes the conflicting interests of Merchants (commercial interests) and Guardians (military/national security interests) in the development of space systems. Security for computer systems presents a similar conflict of commercial and security interests. While computer technology was initially developed to support military interests during and following World War II, by the mid-1990s it had become a predominantly commercial enterprise, and the Internet followed suit. Before and during the economic boon of that period a prevailing policy perspective was to let the industry alone and reap the economic benefits of its labors. The information technology industry is now accustomed to a hands-off approach from the government and is resistant to attempts to encourage actions that are seen as contrary to their market interests.

The Strategy tries to deal with this tension by making the federal government a leader in computer security. However, the government no longer has the large market share in computers that it did in the 1950s and 1960s. Leading by example is not likely to sway many companies. The government has indirect policy tools at its disposal that could be used to encourage information technology companies to produce more secure products. They could influence the industry to improve their security products and processes through government purchases of computer

---

[1.] *Merchants and Guardians: Balancing U.S. Interests in Space Commerce.* Washington, D.C.: RAND Corporation, 1999.

equipment (though it's unclear how much market pull the government might have), or through requirements imposed on other institutions doing business with the government. The government could also change product liability laws to shift the burden of security onto the vendors rather than the consumers. It could educate the public about computer security and encourage them to purchase products that have a particular level of security. Tax policy could be utilized to reward businesses that utilize secure computer products.

The current climate regarding the information technology industry in the United States places most of the power to change cybersecurity with industry. The industry lacks sufficient market incentives to change. While a free market is an important part of a capitalist system, and no doubt helped the information technology industry reach its current level of success and maturity, markets tend to downplay non-market concerns. The government, as a guardian of the public trust, should intercede if other interests are at stake that outweigh their impact on the market. While an arguable point in this field, maintaining public trust and ensuring vital services should be sufficient to at least prompt a debate over whether additional encouragement of industry is warranted where computer security is involved.

While Merchants may have the upper hand in the current climate, policy makers should take care to make sure the Guardians are not given too much power. The total security encouraged by Guardians is at best problematic for Merchants who want to use computer systems to conduct their business, and for the consumers who wish to use the Internet for their own reasons. Access controls could be tightened, for example, but access should not be so cumbersome at designated public entry points that it would discourage interest (and by extension, commerce). Access to the company's human resources database or customer lists is another story. Merchants would be doing a disservice to their customers and/or employees if they did not act more like Guardians for some portions of their computer networks.

While both Merchants and Guardians need to temper their interests when it comes to computer security, the Merchants will be a more important constituency when developing policies for cybersecurity. The private sector is the major actor in this field by virtue of how much information technology and related services it produces and uses. This is also true for critical infrastructures, where industry controls a lot of the structures and organizations involved. Government can take the lead in encouraging action, increasing awareness and expanding the knowledge base, but without cooperation or acceptance from the private sector, their efforts will not be very effective. Much like software vendors depend on users installing patches in order to make their systems more secure, government will have to depend on the private sector to make sure that the monitoring and detection capabilities it develops are used, and that needed information is shared between parties. Otherwise computer security will continue to resemble the Maginot Line and be a brittle defense, shattering at the first concerted attack on a known or suspected weakness.

A Process, Not a Product

Humans are an important element of any type of security. Human security guards monitor the entrances to many office buildings and watch video monitors of other areas. Monitoring can be automated to some extent (and can be essential given the speed of computer traffic), but a human perspective is helpful. Bruce Schneier, a noted cryptography and computer security expert, places humans at the core of his computer security company's operations. While a computer can be programmed to recognize abnormalities associated with cyber attacks, humans are still better equipped to notice the odd things, the unexpected and the unusual. At least until computers develop intuition.

The human element need not be limited to security guards or computer monitors. An important tool of hackers (or any successful con artist) is social engineering. The writings of both Bruce Schneier and Kevin Mitnick emphasize the relative ease by which important information can be gathered by people simply asking the right questions and being really nice to people. People want to be helpful, especially if that's an important part of their jobs. Unless trained to do so, they won't think to verify a person's identity or give a second thought to whether they really need or should have this information. The same is true in espionage. While knowledge and information can be gathered through electronic eavesdropping such as wardriving (driving around with a wireless-enabled computer to attempt to gain access to wireless networks—a practice that demonstrates how insecure many wireless networks are), often the most damaging acts of espionage involved turning an employee to provide information to the enemy.

The need for education and awareness of computer security is mentioned in many government policies and documents concerning computer security. The I3P emphasized legal, social, and economic issues in its research and development agenda for cybersecurity. But how can a government policy effectively encourage education and training for individuals to make themselves more security savvy? The Common Sense Guide issued by the Internet Security Alliance is a good start for raising awareness and providing guidance. It is incredibly difficult for government policies to directly change individual behavior outside of making noncompliance punishable through criminal or civil action. The scope of behaviors involved in cybersecurity makes that practically impossible, even if it were seen as necessary and reasonable.

Policy mechanisms that successfully change human behavior usually do so indirectly. Training all employees to recognize social engineering and be more cybersecurity savvy will require indirect measures such as making it part of security standards, or as a condition of doing business with the government. Changing insurance laws could be another means to require needed training, though it would be easier, if not quicker, for the industry to adjust its policies regarding computer related losses, as it has started to do. Lawsuits can motivate changes in behavior as well. Policies can also change the architecture, or code, of a system to ensure a change in actions. For example, passwords should not be common words, or all letters or all numbers, in order for them to be harder to crack. But some people do not follow such guidelines. Computer systems can be changed at the architecture or code level to enforce these guidelines. Of course, that may prompt people to put their password on a scrap of paper near their terminal, but changes in code can influence behavior.

Ultimately, cybersecurity must struggle with the risks embedded within computer systems. Defending against cyberattacks and other security risks relies on effective human awareness and participation. In computer networks it's not just the technology that matters, but also the people that use it. Policies and strategies must be crafted with considerations of use in mind. If cybersecurity policies are implemented that people are not willing to follow, then the policies are worse than worthless. If management feels the policies make them more secure, noncompliance by employees contributes to a failure of security and a false sense of security—a potentially catastrophic combination. Bruce Schneier describes security as a process, not a product. Security policy is not a single policy, but a collection of policies. It is not just tactics, but also strategies. Security must be integrated into the organization that uses the technology just like it must be integrated into the technology they use. People are critical to the success or failure of security. As much as it might seem the other way around, people use technology. A security product or policy is only successful if people are consistent and in implementing it.

Conclusion

Following September 2001, security is on the minds of Americans a lot more than it used to be. The federal government is spending more money on protecting the domestic front and otherwise protecting its citizens from the risks of an uncertain world. Many openly question how effectively the country can increase security when American society is so open and porous. In that respect, computer security is no different from homeland security. The Internet is open and porous as well. Several things can be lost through increased security measures. Whether it's longer lines at airports or slower connection times to e-commerce sites because they verify your identity, security often requires sacrifices. Having too much security will diminish the robustness and flexibility that makes computer systems and American society so attractive. These tradeoffs need to be debated at many levels of politics and policy, and policy makers need to be particularly careful about how policies regarding computer security influence other policies, and are influenced by policies in other areas (such as economic support for the information technology industry). The policy system can be as complex as the computer systems it tries to influence.

In cybersecurity, the federal government is just one of several players in shaping and implementing policy. Many people, institutions, and other organizations can do their part to make computer systems more secure and encourage others to do the same. No single policy is going to fit all, but rather a portfolio of policies, aimed at many different targets and using a multitude of means is required for an effective strategy to increase cybersecurity. With the ever-changing character of the technology and the networks it comprises, such strategies and policies need to be continually monitored and revised, much like the computer activity that conceals and reveals threats to cybersecurity. In such a rich landscape of players, policies, institutions, and technology, it is clear that a successful approach to security requires careful management of the tension between protection and interaction. To paraphrase the old Russian saying, you must trust and verify to make the best use of the potential of computer networks.

Further Reading

Alter, Steven. *Information Systems: The Foundation of E-Business* (4th Edition). New York: Prentice Hall, 2001.

Mann, Charles C. "Homeland Insecurity." *The Atlantic* 290:2 (September 2002), pp. 81–102.

Mitnick, Kevin and William L. Simon. *The Art of Deception: Controlling the Human Element of Security.* Indianapolis, Indiana: Wiley Publishing, 2002.

National Research Council. *Trust in Cyberspace.* Washington, D.C.: National Academies Press, 1999.

National Research Council. *Cybersecurity Today and Tomorrow: Pay Now or Pay Later.* Washington, D.C.: National Academies Press, 2002.

National Research Council. *Making the Nation Safer: The Role of Science and Technology in Countering Terrorism.* Washington, D.C.: National Academies Press, 2002. (Chapter 5 deals with information technology).

President's Critical Infrastructure Protection Board, *The National Strategy to Secure Cyberspace.* Final (February 2003) draft available at *http://www.whitehouse.gov/pcipb/*. The November 2002 draft can be found at *http://ftp.pcworld.com/pub/new/privacy_security/cyberspacedocument.pdf*.

Schneier, Bruce. *Secrets and Lies: Digital Security in a Networked World.* New York: Wiley and Sons, 2000.

Wulf, Wm. A. and Anita Jones. "Cybersecurity," *The Bridge* 32:1 (Spring 2002), pp. 41–45.

PBS Frontline Program, "CyberWar!" website (view the program and several other features): *http://www.pbs.org/wgbh/pages/frontline/shows/cyberwar/view/*.

Chapter 6

Privacy: Erosion or Evolution?

Najma Yousefi

"Orwell cautions us against 'Big Brother.' Perhaps a more uncontrollable threat is the 'Little Brother' next door."

—George B. Trubow

The year 2000 had mixed promises for the information technology industry. While the worries of Y2K bug faded away, a new challenge came to unfold as several lawsuits were brought before state and federal courts against online service providers for their alleged violations of users' privacy rights. America Online (AOL) was accused that version 5.0 of its AOL software, once loaded on a computer system, caused such changes to the system configuration that would conflict with any non-AOL communication software. Class actions were filed against Amazon.com for its unauthorized collection of Internet users' data by employing its Alexa software, which Amazon had begun to use following its acquisition of

Alexa Internet, Inc. in 1999. Yahoo! Inc. and its affiliate, Broadcast.com, also were sued in Texas for their violation of Texas anti-stalking law on the grounds that, plaintiffs alleged, cookies used by Yahoo! monitored and stalked users.

In yet another lengthy and complex privacy litigation beginning in January 2000, complaints were filed in federal and state courts against DoubleClick Inc., and were subsequently consolidated into a class action. A publicly traded company and the world's largest provider of Internet-based advertising products and services, DoubleClick functioned as a third-party provider of ad services by displaying online ads for over 11,000 websites, 1,500 of which were among the world's most highly trafficked sites. Using its proprietary DART technology, DoubleClick placed on users' hard drives cookie files that collected users' data, hence creating profiles of Internet users and customizing ads for specific demographics. DoubleClick was alleged to have used personally identifiable information (PII) for the purpose of profiling Internet users' browsing behavior—a practice that had ramifications for privacy rights.

What evoked suspicion about DoubleClick's use of PII was its merger in June 1999 with Abacus Direct Corporation, a direct-marketing services company that had maintained a database of personal information on approximately 90 percent of American households. Soon after the merger, DoubleClick formed the Abacus Online Alliance and amended its privacy policy by removing the assurance that it would not combine users' online information with their off-line PII. An allegation was made that DoubleClick had planned to integrate Abacus's offline customer's database with its consumers' online profiles. In a Senate hearing held in July 1999, Marc Rotenberg, executive director of Electronic Privacy Information Center (EPIC) had voiced grave concern at DoubleClick's business practices, and especially its merger with Abacus. Although DoubleClick's CEO announced in March 2000 that they were not pursuing a plan for the integration of the two companies' databases, there were concerns, based on a February 2000 statement by DoubleClick's president, that they had already merged some 50,000 to 100,000 records from the two databases. Meanwhile, the Federal Trade Commission (FTC) launched an investigation into the use by DoubleClick of consumers' information, including the integration of the companies' databases, to determine whether such activities constituted unfair or deceptive trade practices.

DoubleClick filed a motion to dismiss the case, contending that the cookies placed on users' computers were only "of or intended for" the first-party websites that used DoubleClick's ad services. The cookies would not collect any information if the users browsed non-DoubleClick-affiliated websites; nor would Double-Click's cookies collect any PII and therefore no intrusion of privacy had taken place. Moreover, users could simply opt out of the placement of any cookies on their computers. DoubleClick's access to the users' computers, it further argued, was a necessary part of fulfilling its service agreement and was explicitly sanctioned by one of the exceptions of the Electronic Communication Privacy Act (ECPA). DoubleClick conceded, however, that its access to plaintiffs' computers had been unauthorized under the Computer Fraud and Abuse Act (CFAA), but it went on to argue that the plaintiffs had failed to plead the damages established under the Act.

DoubleClick's motion was granted. The court did not find DoubleClick to have violated privacy laws as cited in the plaintiffs' complaints. The court ruled that, despite DoubleClick's admission to have had unauthorized access to plaintiffs' computers under the CFAA, collection of demographic information was a normal practice by marketers and the value of such information to marketers did not constitute damage to consumers. Subsequently in January 2001, the FTC concluded its investigation and announced that it had not found DoubleClick to be engaged in unfair or deceptive trade practices.

Neither the plaintiffs nor privacy advocacy groups were pleased with the verdict. Although DoubleClick's practice did not constitute an actionable conduct under existing privacy laws, it remained an open question as to whether Double-Click had violated Internet users' privacy through unauthorized placement of cookies on their computers and by monitoring their surfing habits. In the above mentioned Senate hearing of July 1999, EPIC's executive director had stated that Internet users had not dealt with DoubleClick, and yet they were targeted with DoubleClick's ads and their information was collected. True, DoubleClick had posted its privacy policy, which informed users of its business practices, including use of cookies and collection of consumers' non-PII data; but the average user was hardly familiar with such policies that are usually permeated with legal and technology jargons. Besides, the long-accepted principles of fair information practices, which have since the 1970s been incorporated into a number of public laws, prohibited use of consumers' data for purposes different from what they were initially intended. It is dubious that DoubleClick's intention for collecting data was made clear to the users and if data were used for their initial purpose, whatever that might have been. DoubleClick had unobtrusive access to private domain, if not private information, and had used the users' computers to send communications to its servers requesting that banner advertisements be sent to the users' computers while the users had no knowledge that their computer systems were being used for such purposes.

While grass-root organizations such as EPIC and Electronic Frontier Foundation (EFF) continued their efforts to convince the court that DoubleClick was guilty of privacy violations, DoubleClick adopted various measures—such as self-regulation, clarification of privacy policy, hiring a topnotch chief privacy officer, and partnering with a privacy consulting firm—to signal its commitment to protection of Internet users' privacy. Nevertheless, in June 2001, a California Judge ruled that a trial would be held unless DoubleClick reached a settlement with the plaintiffs.

In March 2002, DoubleClick reached a settlement agreement with the plaintiffs and, while admitting no wrongdoing, committed to principles of fair information practices and protection of online privacy. As part of the agreement, which was approved by the federal court in May 2002, DoubleClick agreed to the following: pay $1.8 million in attorneys' fees; post an easy-to-read privacy policy to its websites; purge consumers' data on a regular basis; adopt an opt-in policy which clearly explains how DoubleClick uses consumers' data and gives users a choice to agree to those methods by clicking on a Yes box; and launch 300 million Internet ads targeted at educating consumers about use of cookies and how their privacy can be protected. Subsequently, in August 2002, a separate settlement agreement was executed between DoubleClick and Attorneys General of ten states. The agreement describes DoubleClick's business practices and its use of non-PII as well as its lack of intention to merge online non-PII with offline PII. DoubleClick also agreed to hire an independent third-party firm to review its compliance with the agreement for the next four years. It further agreed to pay the states the amount of $450,000 to cover their costs, such as investigation and consumer education. The states ensured to include a clause that the agreement was not to be construed as their approval of DoubleClick's business practices.

Privacy groups, such as EPIC and Junkbusters, protested the court's approval of the settlement, contending that the agreement did not address the class members' interests. They believed that DoubleClick should have been punished for its violations of privacy rights. Nevertheless, the proposed settlement agreement was finally approved by the court on May 21, 2002. It seems as if the settlement agreement was all that the plaintiffs could gain after such a lengthy litigation; this is, perhaps, because it was not clear whether any violations could ever be substantiated under existing laws and, if so, what kinds of remedy could be available to the plaintiffs. Absent clear-cut regulations, it was too costly for both the defendant and the plaintiffs to further pursue the suit.

The DoubleClick litigation epitomizes the complexities involved in the impact of ubiquitous data processing and communication technology on personal privacy. Once one enters the network of a credit card company or a direct mailing group, it will become virtually impossible to find out how, and for what purposes, the personal information will be used. These concerns are not limited to online services as the personal data collected offline, such as filling a paper-based application or warranty registration form for home appliances, are input into mammoth interorganizational networks as well. The burgeoning online communication has also created a market for surveillance technology, which enables commercial entities, as well as individuals, to monitor surfing habits of Internet users. The widespread use of spyware (previously referred to as cookie) has given rise to the emergence of the so-called snoopware technology, which can be installed remotely on a computer system to monitor the user's activities other than Internet surfing habits. Government, too, has long used surveillance technology and, in light of the ongoing war on terrorism, currently has certain projects, such as the Department of Defense's Terrorist (formerly Total) Information Awareness (TIA) program, underway to create the next-generation surveillance technology.

The United States Constitution and the Bill of Rights have established basic individual rights. Privacy rights also have long been recognized in the American common law and statutes, however, there are concerns that privacy laws, as well as the predominant self-regulatory approach, which is rooted in a laissez-faire political philosophy, have been unable to catch up with the rapid growth of technology. There are doubts about the effectiveness of those laws in addressing privacy concerns that are peculiar to communication and data processing technologies. This new dimension of privacy concerns—broadly referred to as Electronic or Information Privacy—has been in the heart of public debate ever since these technologies first found their way to government and commercial organizations in the mid-1950s. Although several privacy laws have since been enacted, privacy activists and social commentators have remained skeptical about the adequacy of predominant sectoral approach to privacy legislation. On the other hand, it is unclear that an omnibus legislation would be capable of encompassing all the complexities of multifarious information technology while at the same time ensuring that other public priorities are not compromised. The interaction of government and other players has thus far failed to create a framework conducive to effective policy making.

This complex situation is the focus of the present chapter.

What Is Privacy?

There exists no universally accepted definition of privacy. Privacy is an amorphous concept with no concrete referent. This explains, for the most part, the lack of consensus as to what may or may not be considered private, and how privacy must be protected. Privacy may be defined in such a way that, relative to a social/cultural context, it would apply to certain domains of personal life where one has expectations of privacy, whatever the term may mean. There are myriad areas of human life which different cultures consider private. Even within the same culture, there is not always unanimous agreement as to what must be regarded as private and therefore be protected from public access. It is not clear whether privacy has solely to do with personal information, or with private spaces, or with the mere behavior of an alleged intruder. Historically, the protection of privacy in the United States has been associated with personal information, but most, if not all, people consider snooping or any intrusive behavior as a violation of their privacy rights regardless of whether any personal information has been acquired. Thus, the diversity of private domains of life rules out the practicality and suitability of a single, comprehensive definition of privacy. Additionally, the elusive character of privacy explains why opinion polls cannot properly reflect privacy concerns; people have different conceptions of privacy in their minds when answering the poll questions. The vagueness of privacy is also responsible for the lack of a coherent public awareness and action in defense of privacy rights.

Since John Locke (1632–1704), the British philosopher, wrote his influential Treatises on Civil Government, certain rights such as those to liberty, privacy, and property, have been regarded as natural rights. Accordingly, society does not grant privacy rights to an individual. This is a normative approach, which posits an inherent value in privacy. Judge Thomas Cooley's legal treatise of 1880 on torts with its much-quoted definition of privacy as "the right to be let alone," and its further expansion in a 1890 classic article by Samuel Warren and Louis Brandeis presume a natural right embedded in privacy. It is noteworthy how Warren and Brandeis warned against the impact of technological advances on personal privacy. There have also been attempts to reduce privacy to such rights as autonomy and liberty. The vast majority of privacy activists are of the opinion that privacy is a fundamental right, which may override other rights, because no other rights can exist in the absence of privacy. This approach relies heavily on legislation to draw the boundaries of personal privacy and on government to enforce such legislation; such a weighty task may not be left to the markets or to self-regulation. Privacy laws, from an activist point of view, must foremost take into account citizens' rights to due process. The European Privacy Directive reflects such a broad-based legal approach.

The rights-oriented approach to privacy has been strongly challenged from an economic point of view, which echoes the ethos of modern society where there is an opportunity cost associated with every piece of information. Private information, from the viewpoint of the economic analysis, is a commodity that can be given up in return for certain goods and/or services. One may opt to protect his/her privacy to the fullest, but he/she may no longer expect services that have to be customized for different tastes—services that have different price tags.

According to the argument, producers of goods and services need to target their consumers, but this is only possible when they have sufficient information about their potential consumers. This approach is rooted in a utilitarian philosophy that attempts to balance the costs of revealing personal information against its economic and/or social benefits. One such approach is adopted in an influential article by the economist Richard Posner (*An Economic Theory of Privacy*, 1978). Just as there are costs involved in concealing facts about oneself, so too are there costs for service providers to discover facts about individuals in order to avert future losses. A job applicant, for example, has incentive not to reveal information about poor health whereas the employer has to ensure that the prospective employee meets the health requirements of the job. Individuals, Posner argues, must not be assisted, via law or other avenues, to conceal facts, which can impose costs on other parties and adversely affect efficiency. This analysis may be construed as an anti-privacy approach in that it ignores the inherent value of privacy cherished by society. Privacy activists have blamed the rapid erosion of privacy on its commodification whereby privacy becomes a tradable commodity.

There is anecdotal evidence, however, that people willingly supply their personal information where they have expectations for desired services.

Privacy concerns emanate from the tension between privacy as a right and privacy as a commodity. Whereas the former requires unconditional protection of the right to privacy, the latter conceives conditions under which privacy as a commodity may (or should) be given up for other interests. The interplay of these two approaches to privacy has spurred advances of communication technologies. Intrusive technology has enabled organizations and individuals to amass information for commercial, political, or personal interests. On the other hand, the increasing concerns about the far-reaching impact of intrusive technology have called forth the emergence of privacy-enhancing technologies (PETs), which limit other parties' abilities to invade privacy by empowering individuals to control the collection and dissemination of their personal data.

However conflicting the foregoing approaches, they underscore personal information as the most concrete aspect of privacy; this can establish a common ground to address privacy issues. The importance of personal information stems from its ability to reflect personal identity. Therefore, the divulgence of one's identity is considered a violation of privacy, for it not only ignores the individual's right to anonymity, it also puts him/her at the risk of identity theft or other potential dangers thereof. The approaches discussed here constitute the core of privacy discourse and contribute to prospective solutions. More important, identification of these conflicting views can help better understand the development of information technology and identify the players who have shaped privacy discourse and policy since the mid-1950s.

Historical Background

When information processing and storage technologies were first introduced in government and private organizations in the 1950s, their far-reaching consequences for civil liberties began to loom large, raising a host of legal, social, and political issues. As the computerization of organizational files facilitated processing and exchange of personal information, stakes were raised for the right of privacy.

The early phase of computerization involved no more than automation of isolated clerical tasks (known as "housekeeping" automation) in a limited number of government agencies and private organizations. The computer revolutionized techniques of organizational management and promised increased productivity and efficiency although the costs of computerization sometimes outweighed its immediate benefits. The use of computers in this early phase, albeit a substantial long-term efficiency gain, did not go beyond normal record-keeping that had hitherto been done manually. Thanks to the callowness of such automation in the mid-1950s, it did not raise significant civil liberties issues.

Further development in organizational data processing is marked by the technological advancement of storage and processing capabilities, which brought about the evolution of data processing beyond its infancy and led to the automation of organizational files. The acquisition in the late 1950s and early 1960s of new computer technologies by some government agencies and large corporations epitomized an epoch-making undertaking that tremendously facilitated the automation of high-activity files on personal data. File automation, however, did not lead to immediate integration in organizations of various records of people, such as driving records and vehicle registration at the Department of Motor Vehicles, or saving and checking account information in banks. By the mid-1960s, file automation paved the way to the emergence of databanks, which, among other capabilities, integrated files on people and immensely helped departments within an organization share information on individuals. IBM's LOGIC (LOcal Government Information Control) was one of the most popular databanks of the 1960s. When implemented in a six-year project (1965–1971) in Santa Clara County, California, LOGIC allowed for the integration of data for ten different county departments and promised significant long-term savings in processing and maintenance of the jurisdiction-wide information. Like other databanks, however, LOGIC evoked debates over privacy and other civil liberties rights at both the local and national levels.

The general fear about databanks was that they would render greater collection of private data possible, making such data available to a larger number of people both inside and outside organizations. The prevailing perception was that the spread of computers in organizations would cause insidious damage to due process by minimizing human input in decision-making (e.g., in an application for loan, or life insurance, or unemployment benefits), thus judging people based solely on codes and numbers giving them little chance to contest decisions made about them. Numerous press articles across the country fanned the flames of this fear and such books as Vance Packard's *The Naked Society* (1964), Myron Brenton's *The Privacy Invaders* (1964), Alan Westin's *Privacy and Freedom* (1967) and Arthur Miller's *The Assault on Privacy: Computers, Data Banks and Dossiers* (1971) became increasingly popular.

A number of research projects were devoted to investigating the impact of computerization on civil liberties. Among them, *Databanks in a Free Society: Computers, Record-Keeping and Privacy* (1972), a research project carried out in the late 1960s under the auspices of the Russell Sage Foundation and the National Academy of Sciences, is one of the best. The study set out to delve into the effects of computerization in 55 government agencies and private organizations, and found that no databank, in its strict technical sense, was in operation at the time. The fear of privacy intrusion, the study demonstrated, was unfounded since no real changes had been made to data processing because of computerization.

According to the study, organizations did not collect more information after launching their automation projects, nor did they make the data available to a greater number of individuals and organizations. More important, organizations continued to maintain sensitive information in manual files and applied the same rules of privacy and confidentiality to computerized data as had hitherto been applied to the manual, paper-based data. Although organizations, especially government agencies, had historically exchanged information, the emergence of computers did not necessarily increase, though certainly facilitated, data sharing.

Privacy concerns were nevertheless on the rise during the 1960s and 1970s even though a real databank had not yet been launched by the early 1970s. The Louis Harris survey on privacy revealed that in 1978, 64 percent of Americans were "greatly concerned about threats to their personal privacy," and one out of three Americans thought that the society was "very close to or already like the type of society described by George Orwell in his book 1984." In the same poll, 54 percent expressed concerns that their personal privacy was threatened by the use of computers. The poll also found an increase from 48 percent in 1974 to 76 percent in 1978 in public concern about the erosion of privacy because of the widespread collection of personal information for commercial purposes (*The Dimensions of Privacy: A National Opinion Research Survey of Attitudes Toward Privacy.* p. 5; New York: 1981).

The fear of a dossier society had in large part to do with a controversial plan, proposed in 1965 by the Social Science Research Council (SSRC) for the establishment of a Federal Data Center to improve the quality of national statistical surveys. Such a federal databank was perceived to put in jeopardy the most essential underpinning of a democratic society, that is, freedom of speech, for citizens would be fearful of potential consequences of expressing their opinions. The public outcry about the National Data Center was great enough to elicit Congressional debates in the 1960s and early 1970s. In July 1966, the House Special Subcommittee on Invasion of Privacy, established in 1964, held the very first hearings in order to investigate the nature and goals of the Federal Data Center and its impact on personal privacy and civil rights. Subsequently, in March 1967, the Senate Subcommittee on Administrative Practices and Procedures conducted its first hearings on this issue and through a questionnaire, sent to all federal agencies, carried out a detailed analysis on the magnitude and accuracy of personal data in federal government. The survey unearthed the existence of 31 billion individual-person records with some degree of inaccuracy and rather weak confidentiality provisions. Despite Congress's unfavorable reaction to the Federal Data Center proposal, the Bureau of Budget insisted on the establishment of such a databank. More congressional hearings ensued. Following a series of hearings in 1967 and 1968, both the House and the Senate rejected the proposal outright. They contended that they had no guarantee that the federal databank would operate within the confines of its statistical mission while observing privacy rights. But the fear of a dossier society with a centralized data center lingered for years to come.

The sense of insecurity vis-à-vis databanks also had to do with the dwindling trust in government that the American society experienced in the post-World War II era. Statistics show that trust in the American government dropped from 71 percent in 1958 to 62 percent in 1962 and then to 37 percent in 1970 (*Databanks in a Free Society*, p. 343). This lack of trust manifested itself, inter alia, in antiwar demonstrations and the civil right movement. Since the U.S. Army and FBI were already collecting detailed information about political dissidents, no matter how lawful their activities were, it was quite natural that the plan for launching a national databank be viewed as policing the society. Although the plan for a federal databank never panned out, the media and social commentators remained distraught about the latent dangers that the emerging database technology posed to civil liberties, especially to privacy.

The ire expressed in the media had to do with the onerous consequences of launching databanks. Security was particularly problematic: if, as it were, a great deal of information was centralized in a small space such as a magnetic tape, as opposed to a huge file cabinet, the risk of theft or unauthorized access was greater, so the public needed assurance of due diligence given to the security of information. The public opinion reflected a widespread sense that file automation, together with centralization and sharing of data, was inherently inimical to privacy.

Public opinion polls suggest that, despite the consensus on the urgency of protecting privacy, people recognized the right of society to have a reasonable amount of information about individuals who expect certain services from public or private organizations. An insurance company, for example, would be unable to offer policy without first being able to asses the risks involved in the insured subject (e.g., life, property, vehicle, etc.). The question, however, was how much information had to be collected and how it was used in rendering service to citizens. Absent certain legal protections, the public needed assurance that personal data were handled with due regard to the rules of confidentiality and, more important, how and under what circumstances they might be shared with private or government organizations.

It is against this backdrop that the notion of "balance" was developed in the early 1970s, and has since guided the privacy debate. While few people may ignore the importance of balance between an individual's right to privacy and society's right to know, it remains unclear as to where the line between the two rights should be drawn. The rapid evolution of data storage and processing technologies have engendered two phenomena that further complicate privacy concerns: first, the advent of the Internet together with such capabilities as online communication and e-commerce; and second, the availability of surveillance technology to individuals and private businesses alike, and its ubiquitous use in daily life, which has changed the landscapes of both physical and virtual spaces. Everyday the media reveal horror stories about the invasion of privacy with the usage of new technologies. A recent report by the Annenberg Public Policy Center of the University of Pennsylvania, Americans and Online Privacy: *The System is Broken* (June 2003), uncovered the dark side of prevalent business practices that have implications for Internet users' privacy. The report showed that the online industry is using cutting-edge techniques to amass users' data and to carry out profiling for marketing purposes. It further revealed that the vast majority of American users of the Internet have very little knowledge and simplistic understanding about their privacy while on the Internet. As the growth of information and communication technologies is gaining momentum and creates tools that once challenged human imagination, diverse applications of technology to personal information is causing transmogrification of privacy as both a notion and a value. It is, therefore, critical to identify the players whose roles influence the evolution of both the technology and the privacy discourse.

Who Are the Players?

Privacy concerns are shaped by the interaction of a number of players who represent different views, values, and interests. Yet, the elusive conception of privacy has blurred the identity of the players that crisscross in a complex nexus of social interactions where multiple roles may be played by each actor, hence increasing the risk of conflict of interest. This has in turn contributed to the growing complexity of the privacy discourse and the search for viable solutions.

Government

Compared with other branches of the American government, the legislative branch has been more active in privacy-related issues. As early as 1965, Congress has held hearings and has produced extensive reports on different aspects of privacy issues. The Special Subcommittee on Invasion of Privacy of the House Committee on Government Operations held hearings from 1965 onward on such privacy-related issues as use of electronic surveillance by federal and commercial entities, the Post Office surveillance of mail and of its employees, psychological testing and polygraph screening of federal employees. The Senate Subcommittee on Administrative Practice and Procedures, and the Subcommittee on Constitutional Rights of the Judiciary Committee conducted hearings on similar topics since 1965. The congressional hearings during 1966–1967 on the controversial Federal Data Center were the most salient manifestation of Congress's concern over the latent threats that the new information technology posed to privacy. Congress produced several reports in this early period and recognized the need for protection of privacy.

Though congressional hearings and reports were effective enough to halt the plan for the Federal Data Center, they invariably failed to create a consensus conducive to enactment of a comprehensive privacy law. The proposed Right of Privacy Act of 1967 was the first Senate bill that addressed threats posed by databanks and the widespread use of personal data by government; the bill never reached a floor vote, however. Such congressional bills that have sought to establish a comprehensive legal/policy framework for privacy have thus far been stopped at the committee level. Nevertheless, the close cooperation of social commentators and privacy activists with Congress has resulted in a series of privacy laws pertaining to different socioeconomic sectors. As such, a sectoral approach has in effect dominated the policymaking of privacy.

The 1970 Fair Credit Reporting Act is the very first of such laws that were enacted in the wake of the 1968–1969 congressional hearings. The Act sought to hold credit bureaus responsible for their handling of personal data which would have direct effect on people's lives. It, too, established standards of storage and maintenance of personal data, and provided mechanisms that would allow individuals to access or correct their personal information. The Act embodies the first official undertaking by government to safeguard privacy by making a potent law. Yet, the Fair Credit Reporting Act of 1970 has been criticized for excessive influence exerted by the credit bureaus' lobbyists, which may have lowered the bar for the protection of private information. The credit industry has reportedly had a direct input in drafting the Act, which might have made the Act more suitable to their preferred practices.

The 1971 Senate hearings on federal databanks conducted by Senator Sam Ervin, Jr., Chairman of Subcommittee on Constitutional Rights of the Senate Committee on the Judiciary, were another important undertaking by Congress. A large number of experts and officials, including Alan Westin and Arthur Miller as well as representatives from the Army, FBI, and commercial organizations, testified before the Subcommittee. Some of the testimonies revealed that, contrary to the aforementioned findings of the National Academy of Sciences' study on databanks, the magnitude of information collected after the computerization projects had in fact increased. It, too, was revealed, based on the testimony by Robert Froehlke, then Assistant Secretary of Defense, that the White House and the Department of Justice had played a chief role in planning the Army surveillance of political dissidents. There was clearly a conflict of interests between branches of government, that is, Congress's desire to curtail the ongoing surveillance of citizens on the one hand, and the Nixon Administration's efforts to unleash the power of information technology for a full-fledged collection and maintenance of dossiers, on the other.

The early years of the 1970s were witness to events that tipped the balance of power in favor of privacy protection. In 1972, Elliot Richardson, then Secretary of the Department of Health, Education, and Welfare (HEW), established the Secretary's Advisory Committee on Automated Personal Data Systems in order to overhaul the impact of data processing and storage technologies on personal privacy. The HEW's report was published in July 1973 and proved one of the most influential privacy reports of all times. Records, Computers, and the Rights of Citizens exhibits an in-depth and detailed analysis of the emerging data processing practices in both government and the private sector. The HEW's report demonstrated that real changes had in fact been made to the magnitude and complexity of collection, maintenance, and dissemination of citizens' personal data. Recognizing limited protection afforded by law at the time, the HEW's report made recommendations, including a Code of Fair Information Practices, which laid the groundwork for the future legislations as well as for the self-regulatory measures (see Box 1). Meanwhile, the Watergate scandal left little room for the Nixon Administration to have reservations about the protection of private information. In his radio address of February 23, 1974 entitled "the American Right of Privacy," president Richard Nixon stressed the fact that the computer, albeit a necessary tool for modern societies, was changing the long-cherished value of the American society (i.e., personal privacy). "Advanced technology," Nixon stated, "has created new opportunities for America as a nation, but it has also created the possibility for new abuses of the individual American citizen. Adequate safeguards must always stand watch so that man remains the master—and never becomes the victim—of the computer." Nixon went on to announce the establishment in the White House of "a top priority Domestic Council Committee on the Right of privacy." Chaired by then Vice President Gerald Ford, the Committee was tasked with the investigation of "three key areas of concern: collection, storage, and use of personal data." The creation of this Committee marks the beginning of the executive branch's direct involvement in the privacy policymaking.

The Committee's close working relationship with Congress played a key role in the forthcoming legislations. Whereas the House and the Senate, relying on the solid foundation laid by the HEW's report, were preparing two separate bills which resulted in the Privacy Act of 1974, the Ford Administration, together with the Committee, strove to offset Congress' pro-privacy agenda with modifications to the bills that addressed certain reservations of both the Administration and the industry. In his statement of October 9, 1974 on privacy legislation, president Ford made it explicit that the right of privacy in legislation "must be balanced against equally valid public interests in freedom of information, national defense, foreign policy, law enforcement, and in a high quality and trustworthy Federal work force" (emphasis added). He went on to urge amending the bill to permit such exemptions that can accommodate those interests. The Ford Administration managed to change a provision of the Act that required creation of a Privacy Commission as a federal oversight agency. Rather, the Office of Management and Budget (OMB) was charged with the responsibility to oversee the implementation of the Act. Over the past four decades, however, very few regulatory or executive guidelines have been issued by the OMB. In December 1974, the Privacy Act passed the House and the Senate vote, but a subsequent amendment proposed by

Box 1: Principles of Fair Information Practices

- There must be no personal data record-keeping systems whose very existence is secret.

- There must be a way for an individual to find out what information about him is in a record and how it is used.

- There must be a way for an individual to prevent information about him obtained for one purpose from being used or made available for other purposes without his consent.

- There must be a way for an individual to correct or amend a record of identifiable information about him.

- Any organization creating, maintaining, using, or disseminating records of identifiable personal data must assure the reliability of the data for their intended use and must take reasonable precautions to prevent misuses of the data.

(Source: *Records, Computers, and the Rights of Citizens,* Department of Health, Education, and Welfare, p. 41 (Washington, D.C.: 1973))

Representative Erlenborn made such substantial changes to the Act that dashed the hopes of privacy activists for a potent, comprehensive legal protection of privacy.

The voting pattern of the amendment to the Privacy Act suggests a liberal-conservative dichotomy with the conservatives overwhelmingly supporting the amendment. A similar voting pattern can be found in some other privacy bills, especially in proposed omnibus legislations, with the liberals and privacy activists arguing in favor of a greater protection of privacy while the conservatives and the private sector lobbyists being against such measures. Yet, it might be misleading to superimpose a partisan pattern on privacy legislations. Space does not permit an exhaustive evaluation of congressional privacy bills. Suffice it to say, bipartisan privacy bills have been proposed at the House and the Senate alike. For example, the Privacy Commission Act (H.R. 4049) is one such bipartisan bill that proposes a 17-member privacy commission appointed jointly by the president and the leaders of the two parties at the House. The Commission will be charged with the investigation of privacy issues and will present its reports for a necessary comprehensive legislation. Proposed in 2000, the bill set out to change the predominant sectoral approach by adopting "a holistic approach," a term used by Representative Asa Hutchinson of Arkansas who, together with Representative James Moran of Virginia, introduced the bill. The bill, however, failed to pass the House vote.

At the same time, a bipartisan bill, the Online Privacy Protection Act (S. 809), was introduced in the Senate, but failed to reach the Senate floor for vote.

The Privacy Act of 1974 required the establishment in the Senate of the Privacy Protection Study Commission, which was the first privacy commission in America, although its role was limited to study, rather than implementation and oversight. The Commission was created at the congressional level in lieu of a federal oversight agency which was initially proposed by the Act, and which faced tremendous objection from the Ford Administration and the republican legislators. In 1977, the Commission issued its report, Personal Privacy in an Information Age, which found that the Privacy Act had not been effective to warrant adequate protection against abuses of personal information. The report identified a need for further legislation to enhance the protection of personal data both in government agencies and the private sector. The Commission's report recommended that employers, when using personal data for recruitment purposes, should heed the Code of Fair Information Practices established by the HEW's report. A few privacy commissions have since been founded and have produced reports that recommended a variety of legislative and self-regulatory measures in protection of privacy. But unlike Europe and Canada where privacy commissions play a remarkable regulatory, as well as executive, role and function as a liaison between all concerned parties and players, the American privacy commissions seem to have stagnated in a state of confusion.

Computer System Security and Privacy Advisory Board (CSSPAB) was a privacy commission established in 1987 as a requirement of the Computer Security Act. The CSSPAB consists of twelve technology experts and is charged with the investigation of computer security and privacy as related to unclassified information in federal government, so its mandate does not include information used in the private sector. The Board is responsible to advise the following officials and entities on computer security and privacy issues: the Secretary of Commerce; Director of National Institute of Standard and Technology (NIST); Director of the Office of Management and Budget (OMB); Director of National security Agency (NSA); and relevant congressional committees. In its September 2002 report, the CSSPAB identified specific areas in government privacy policy and management, and made recommendations for improving the current condition of privacy policy. It is not clear as to how these recommendations may transform to concrete decisions and actions. This is in large part due to the lack of leadership in privacy policymaking whereby different congressional committees, government departments, commissions, and advisory boards are partially involved in this important matter.

In 1995, the Information Infrastructure Task Force (IITF) was established at the White House in order to strengthen the federal government's leadership in National Information Infrastructure (NII). Chaired by Ron Brown, then Secretary of Commerce, the IITF was an attempt to bring together different federal agencies, as well as representatives from the industry and academia, involved in information infrastructure. The Working Group on Privacy was then formed under the Information Policy Committee to design the Administration's policy on privacy matters, hence harmonizing privacy policymaking. The formation of the IITF was prompted by the Clinton Administration's awareness of the fact that the rising privacy issues could impede the rapid growth of the Internet commerce. In May 1994, the Working Group issued the first draft of its report, Principles for Providing and Using Personal Information, and subsequently released its second draft in January 1995. Privacy advocacy groups were quick to criticize the content of the report. The Center for Democracy and Technology issued its commentary in March 1995 and stated that the Working Group's report had deviated from the principles of Fair Information Practices set forth in the HEW's 1973 report. Meanwhile, the Department of Commerce, as an IITF member, was asked to investigate the feasibility and suitability of legislating the Internet and its potential impact on the growth of e-commerce. The Department of Commerce's report, *A Framework for Global Electronic Commerce*, issued in July 1997, recognized the importance of protecting online privacy. However, the report recommended that the industry's initiatives in self-regulation, rather than federal legislation, would better protect online privacy. The principles and guidelines developed by the report were again criticized by privacy groups for their lack of attention to the HEW's principles. Yet, it appears that, toward the end of the Clinton Administration, the Working Group on Privacy was considering a more comprehensive, regulatory approach to privacy policy. This is conspicuous in its close working with the FTC whose year 2000 report to Congress highlighted the need for a federal legislation. However, this trend was reversed by the year 2000 presidential election, which resulted in the change of leadership at the FTC.

In February 2000, the U.S. Congress was witness to the formation of its first bipartisan, bicameral Congressional Privacy Caucus (CPC). Co-chaired by members of the two parties at Congress, the CPC has been able to hold hearings on privacy issues and has made great efforts to secure support from other members of Congress for privacy bills. The CPC has been mostly focused on financial privacy, especially on loopholes in the Gramm-Leach-Bliley Financial Services Modernization Act of 1999 (this Act is briefly discussed under Statutory Protection of Privacy). The CPC has been working with consumer groups and privacy activists to address concerns of privacy community, but has had limited success in the passage of new privacy bills. Thanks to the CPC, privacy issues are receiving a greater attention, however, it seems as if the pursuit of a comprehensive statute has been put on the back burner.

The involvement of various government agencies (e.g., Congress, FTC, Departments of Commerce, Justice, Health and Human Services, Working Group on Privacy, etc.) is indicative of government's attention to this matter, yet it portrays the lack of harmony and focus in privacy policymaking since these agencies do not always follow the same agenda. On the other hand, there has been a tension between different interests (i.e., privacy, security, economic prosperity, public services, etc.) that government as a whole aspires to advance. It is a matter of convention as to which value should take precedence; in reality, other priorities more often than not have taken precedence over privacy. Such is the case with the enactment of the USA PATRIOT Act of 2001 that, in the wake of the September 11 terrorist attacks, enhanced government's authority to keep a closer eye on the private affairs of citizens who may be suspected of terrorist activities.

The recent controversy over the Department of Defense's (DOD) Terrorist (formerly Total) Information Awareness (TIA) program further highlights this tension. Designed by the Defense Advanced Research Project Agency (DARPA) and under the leadership of retired Admiral John Poindexter, the former National Security Advisor to President Ronald Reagan and a controversial figure in the Iran-Contra scandal, the TIA program seeks to create advanced technologies to gather and analyze a variety of data for such operational purposes as to deter potential "foreign" terrorist attacks. In light of its ramifications for civil liberties, grass-roots organizations have voiced concerns about the TIA's long-term impact on personal privacy. In February 2003, Congress passed the Data-Mining Moratorium Act (Pub. L. 108-7), which prohibited allocation of budget to the TIA program or any data-mining project by a federal agency until a law is enacted to specifically authorize such projects. The Act required the Secretary of Defense, the Attorney General, and the head of any federal agency that uses or develops data-mining technology to report to Congress on such projects. DARPA's Report to Congress regarding the Terrorism Information Awareness Program, submitted on May 20, 2003, was intended to address Congress's concerns about the TIA's impact on privacy. While recognizing its implications for privacy, the report affirms that the TIA program will operate within the purview of the U.S. law and will heed civil liberties. Changing the name of the project from Total Information Awareness to Terrorism Information Awareness, the report is an attempt to establish that the TIA program is strictly focused on "counterterrorism information architecture" and will not seek to create a centralized database. It further asserts that the TIA program will comply with laws pertaining to information about U.S. citizens.

Despite the DARPA's report to Congress, the Senate's Defense Appropriations bill included an amendment, proposed by Senator Ron Wyden of Oregon, that prohibited further allocation of funds by the DOD to the TIA program or any components thereof. The bill passed the Senate vote on July 17, 2003. While the amendment was pending the passage of the House vote, the Bush Administration opposed the amendment for its imposing of restrictions on government's use of certain tools in the war on terrorism. About two weeks later, the DOD announced its decision to shut down the Policy Analysis Market (POM), a component of the TIA program and a precursor to DOD's Future Market Applied to Prediction (FutureMAP) whose budget was denied by the amendment. The PAM was designed to receive online bids for sale on the possibility of future's major political events, such as wars, coups, assassinations, and terrorist attacks. Subsequently, Admiral Poindexter submitted his resignation as the director of the Total

Information Awareness Office. In an Op-Ed article in the *New York Times* published on September 10, 2003, Admiral Poindexter tried to clarify some of the critical questions raised about the TIA program. Further, he warned that Congress seemingly had the intention to cancel funds needed for the development of innovative antiterrorist technologies.

Congress appeared determined to halt the TIA program when the House/Senate Conference on Defense Appropriations on September 24, 2003 announced Congress's decision to eliminate budget for the Office of Total Information Awareness, including the TIA program. It, however, allowed funds appropriation for other ongoing antiterrorism research programs at DARPA so long as they have no data-mining goal and are lawfully geared toward foreign counter-terrorism intelligence as specified in the Conference report. Regardless of whether or not president Bush will sign the Conference report, the tension between privacy and security will continue to affect government policies concerning information privacy and future legislations.

Service Providers and Technology Manufacturers

Since the 1960s, providers of different types of services (e.g., insurance companies, credit bureaus, banks, online stores, online advertisers, etc.) have exerted influence on privacy policymaking. They have participated in congressional hearings and have worked with government officials to have their interests and concerns addressed in privacy laws. The Associated Credit Bureaus, Inc., for example, reportedly had a direct input to drafting the 1970 Fair Credit Reporting Act; their input, some privacy activists argue, made the resultant law more suitable to the preferred practices of credit bureaus although the Act provides a moderate level of protection against some abuses of citizens' financial information. Likewise, the National Association of Businessmen and the Americans for Constitutional Actions are said to have exerted a direct influence on introducing and the passage of the amendment to the Privacy Act of 1974.

Service providers strive to respond to market demand for their goods and services while maximizing their profit. In light of public concerns about privacy, the market demand for, say, Internet-based services is likely to wane unless service providers can first provide assurance that customers' privacy would not be compromised. Thus, service providers have to disclose their privacy policies in order to gain the public confidence. If, for example, a service provider shares customers' information with their affiliates or third parties, they are required by law to disclose such business practice and to give their customers the option to opt out; (privacy activists have been trying to convince lawmakers to change the law in such a way as to require service providers to obtain customers' express consent by agreeing to opt in). Service providers also need to assure their customers that efficacious security technologies are in place so that online exchange of data over their websites is not subject to intrusion. Implementation of security technology is costly in that the adoption of strict privacy policies is likely to abate some sources of revenue, such as selling customers' information. In fact, the fear of increased costs associated with adopting tighter security and stronger privacy-enhancing measures has put service providers in a position to resist strict privacy laws. Under such circumstances, service providers will more than likely externalize the cost of enhancing privacy by increasing prices to customers in order to cover the cost of security implementation or to compensate for their lost revenues. If, however, such security measures were either required by law or highly demanded by the public awareness, then the prices would be expected to decline over the course of competition.

The industry lobbyists have been reminding policy makers of enormous costs, for both the government and service providers, involved in protecting privacy. During the 1968 House hearings on credit and insurance surveillance, John Spafford, an industry representative, testified that firms represented by him honored information requests from government agencies without asking for a subpoena. He further argued that issuing a subpoena would increase the government operations cost, to which Representative Gallagher responded, "a little extra costs in Government Operations to prolong the civil liberties of people and their constitutional rights is a good investment" (House Committee on Government Operations, Commercial Credit Bureaus, 90th Congress, 2nd session, p. 4. Washington, D.C.: U.S. Government Printing Office, 1968).

In working with Congress, industry representatives have assured lawmakers of their commitment to the protection of privacy. The introduction of the position of chief privacy officer, especially in public companies, is a sign of such commitment. They also have voiced their concerns about the adverse effects of hasty legislations, which ignore the providers' increased operation costs and fail to take into account necessary consistency with previous statutes. In the House hearings of April 2001, An Examination of Existing Federal Statutes Addressing Information Privacy, AT&T's chief privacy officer pointed out inconsistencies of a few federal statutes that directly affect his company's operation. He indicated that "a high standard of privacy" at AT&T was the result of self-regulation. Indeed, the vast majority, if not all, of service providers who deal with customers' personal data are of the opinion that a self-regulatory regime works more efficiently and creates more satisfactory results. From their point of view, legislation imposes mandatory rules that may impact their profit-maximizing engine while increasing the risk of lawsuit in the event those rules are not strictly followed. There is evidence that government has been mindful of this issue. In the same hearings, Representative Edolphus Towns of New York criticized the passage of Children Online Privacy Protection Act (COPPA) for its lack of attention to self-regulation. He also gave his support to piecemeal legislation, which in his opinion is consistent with the history of privacy policymaking in America.

Service providers' concern is arguably a legitimate one, aside from their profit-maximizing motive. Hence, they seem to be reluctant to place a higher priority on privacy protection vis-à-vis cost-effectiveness and efficient operations unless legislation makes privacy-enhancing measures a necessary part of their fixed cost. As mentioned earlier, privacy is a complex concept, which can have different meanings for different socioeconomic sectors. It will be extremely difficult to establish a privacy protection baseline, embodied in an omnibus legislation, that would be equally applicable to all socioeconomic sectors. This is for the most part what has prompted a sectoral approach to privacy in the United States. However, the lack of a holistic approach and a baseline of privacy standards will continue to cause inconsistencies in the overall privacy policymaking. There is not much reason to think that the tension between self-regulation and legislation on the one hand, and between piecemeal and comprehensive legislations, on the other, will subside anytime soon.

Technology manufacturers are in a more complex situation relative to service providers. Whereas part of their demand comes from service providers who have a need for data processing and communication technologies, there, too, is a public demand for PETs. Technology manufacturers are also expected to meet security and privacy standards in their products even though such standards are vaguely, if at all, defined in existing laws.

Three Microsoft products provide an insight into this complex situation. First, Microsoft incorporated certain privacy-enhancing features into its version 6.0 of Internet Explorer. Perhaps the most important feature is the integration of P3P (Platform for Privacy Preference) that enables Internet users to visit only those websites that satisfy the users' privacy preferences. Microsoft was not required by law to integrate such a feature in its Explorer, but it opted to do so in an effort to respond to a market demand for products that are sensitive to users' privacy. Nevertheless, in their joint report (*Pretty Poor Privacy: An Assessment of P3P and Internet Privacy*) in June 2000, EPIC and Junkbusters took on P3P's failure to meet certain privacy standards. As well, in a complaint filed in July 2001 with FTC, EPIC, on behalf of twelve other consumer groups, asserted that Microsoft's adoption of P3P provided little protection of Internet users' privacy (see below for more details on this complaint).

Second, with the introduction of .NET services under the Windows XP umbrella, Microsoft began to increase its share in the growing market of online services. Among other services, Microsoft released its Passport with three related Internet Services: Passport Single Sign-In (Passport); Passport Express Purchase (Wallet); and Kids Passport. Passport services collect customers' PII and allow them to sign in with a single username and password at participating websites in order either to access information or to make a purchase. Passport operates on the Microsoft HailStorm platform which enables the online exchange of information with Microsoft business partners. Following the July 2001 complaint filed by EPIC, the FTC initiated an investigation into the alleged privacy and security flaws of Passport services. Microsoft had failed, the complaint charged, to live up to the security and privacy expectations set in its software agreement while it had led consumers to believe that Passport was protected by reliable security features and a strict privacy policy. Passport services, it further alleged, move the control over PII away from users and make it vulnerable to potential abuse, since Microsoft's privacy policy, as well as privacy requirements that Microsoft set for its partners, does not offer a strong protection. These salient weaknesses, according to the complaint, constituted an unfair and deceptive business practice. In August 2002, Microsoft agreed to settle the charges by committing to refrain from any misrepresentation of its information practices pertaining to Passport and other similar services as well as by implementing a comprehensive security program for such services. Microsoft's decision to settle the charges, however, was announced nearly one year after the release of its Windows XP in October 2001.

Third, since the introduction of its Windows XP, charges have been leveled against Microsoft that its new operating system violates users' privacy rights. The aforementioned complaint by EPIC stated that tracking features known as Digital Rights Management (DRM), incorporated into Windows XP to detect unauthorized use of its software and to crack down on piracy, were a clear violation of privacy rights. The use of such features allegedly provides Microsoft with access to users' computer systems.

Although instructive, the Microsoft case may not apply to all technology manufacturers who do not enjoy the same level of diversity of their products and resources, nor the same type of dominance in their market share. As the IT industry is becoming increasingly segmented with companies investing in different specialty areas, it can be envisaged that the market demand, or requirements set by law, for security and privacy protection would stimulate faster growth of PETs. By the same token, intrusive technology could well continue to grow if law does not set constraints on their applications, or if the public loses its sensitivity to their use.

Non-government Watchdogs

Grass-roots organizations and privacy advocacy groups have long played a crucial role in the politics of privacy. It is worth noting that, unlike the other players discussed above, privacy advocacy groups do not *prima facie* strive for their own interests per se, since they represent citizens' interests. Their role in the policy debate is to make the public aware of its rights. The importance of their role stems from the fact that they can identify burning issues while there is a lack of public consensus about the kinds of behavior that constitute a violation of privacy rights. Social commentators and activists must be credited with making privacy a perennial public policy issue since the early 1960s. They have worked closely together to form a privacy policy community while at the same time providing Congress with immense help in preparing for public hearings. When in 1974 the Republicans in Congress were working with the industry lobbyists on an amendment to the Privacy Act, several privacy groups—especially Americans for Democratic Action, the AFL-CIO Committee on Political Education, and the Consumer Federation of America—joined forces with the Democrats to stop the amendment, to no avail.

American Civil Liberty Union (ACLU) is, perhaps, the most eminent of such groups. Established in 1920, ACLU has been dedicated to the preservation of individual rights under the U.S. Constitution. While ACLU was initially active in such areas as anti-war movement, minority rights, and freedom of speech, it grew to an influential non-partisan organization that took a stance on privacy issues in the face of increased computerization of personal data. With the rise of information privacy issues in the 1960s, ACLU established its Privacy Committee, chaired by Alan Westin, and proceeded to create a Privacy and Technology Project. From the 1960s onward, ACLU has been working with Congress, as well as with other grass-roots organizations, to facilitate the introduction of privacy bills. ACLU has now grown to a non-profit organization with nearly 400,000 members and supporters, and with offices throughout the United States. In 1977, when the HEW proposed Project Match to undertake a massive data matching, consisting of extensive comparison between files of federal employees and files of federal aid recipients, ACLU joined forces with the Senate Privacy Protection Study Commission and a few other advocacy groups to halt the implementation of the Project. Though this effort failed to stop the implementation of Project Match, the Project eventually became bogged down in the processing and analysis of enormous amount of personal data that it had generated.

The Electronic Privacy Information Center (EPIC) is another influential advocacy group that has been specifically focused on privacy issues. Its director, Marc Rotenberg, used to be an intern at ACLU's Privacy and Technology Project. As discussed earlier in the DoubleClick, TIA, and Microsoft cases, EPIC, as a public interest research center and a non-government watchdog, has been raising public awareness about constitutional rights and their violations. Established in 1993, EPIC has been active in the privacy policy debate and a variety of issues germane to electronic information processing and online communication. EPIC has filed numerous lawsuits against public and private organizations for their violations of privacy rights. In addition, EPIC has joined other civil liberties groups as co-counsel or *amicus curiae* (friend of the court). In July 1999, EPIC participated in a Senate hearing challenging a report from the FTC on the Internet Privacy. In its efforts to raise the public awareness, EPIC has published a number of books and reports on privacy rights.

Electronic Frontier Foundation (EFF) is another grass-roots organization that, since 1990, has been involved in a wide array of issues arising from the widespread use of information technology and its impact on civil liberties. EFF works closely with other civil liberties groups and provides legal/technical counseling to both individuals and policymakers. EFF maintains a pro bono legal database on cases associated with civil liberties. Like EPIC, EFF has filed lawsuits against alleged violators of privacy rights. The case of *Steve Jackson Games v. U.S. Secret Services* is one of the first litigations that established the privacy of email. In addition to its various educational programs that aim at enhancing the public knowledge, EFF has since the early 1990s published its biweekly electronic newsletter which has over 27,000 subscribers.

There are many other activist groups who, together with ACLU, EPIC, and EFF, comprise the community of privacy activism. This community includes, among others, U.S. Public Interest Research Group (U.S. PIRG), the Privacy Rights Clearinghouse (PRC), the Online Policy Group (OPG), and Public Citizen.

Protection of Privacy in the United States

The United States legal system offers three main sources for the protection of privacy: constitution; legislation; and common law. The American approach differs from the European method of privacy protection in that it has shown reluctance to legislate a broad-based privacy statute, in much the same way as there is no single agency responsible for enforcing privacy laws. The United States legal system demonstrates a sectoral approach which has enacted or amended privacy laws specific to different socioeconomic sectors and in response to new circumstances, including the emergence of new technologies. This type of reactionary method, however, is consistent with the self-regulatory approach that has dominated the privacy policymaking in the United States.

The following sections offer a brief overview of each source of privacy laws. As the limited space of this chapter does not permit an exhaustive analysis of all laws, we will only touch upon a few cases before the U.S. Supreme Court as well as select federal statutes that have proved influential in the policy debate of information privacy.

Constitutional Protection of Privacy

The United States Constitution and the Bill of Rights do not contain any mention of the term privacy. The Fourth Amendment protects citizens against government's unreasonable searches and seizures while the Fifth Amendment protects them against self incrimination. Even with a liberal interpretation to regard privacy as the *raison d'être* of the Fourth Amendment, there is still huge disagreement as to whether the constitutional protection against government's physical intrusion can be extended to electronic personal data and communications.

Equally important, the constitutional right to privacy, if anything, serves to protect citizens against government agencies' abuse of personal information and their intrusion to private spaces, and therefore cannot be readily extendable to similar practices by private parties.

The rapid growth of technological inventions has since the late nineteenth century posed serious concerns about the impact of new technologies on personal privacy and, more so, about how legal disputes should be settled within the purview of the Constitution. The United States Supreme Court has made numerous important, albeit inconsistent, decisions on privacy cases, which vary from sexual orientation and the use of contraceptives to wiretapping and computerized patient databank. The diversity of these cases, together with the changing nature of the technology, has contributed to the inconsistency of the Court decisions, hence causing grave confusion about the privacy of personal information.

In 1928, the United States Supreme Court made one of its first privacy rulings that involved the use of new technology. In *Olmstead v. United States*, Olmstead and eleven other coconspirators were convicted of illegal trade of liquor between the United States and Canada. The evidence used in this case was obtained by the Seattle Police through intercepting the defendants' office and home telephone lines. In its ruling, the Supreme Court decided that individuals had no reasonable expectation of privacy when using the telephone. The Court ruled that "one who installs in his house a telephone instrument with connecting wires intends to project his voice to those quite outside, and that the wires beyond his house, and the messages while passing them, are not within the protection of the Fourth Amendment" (277 U.S. 438, 48 S.Ct. 564 (Mem), U.S., 1928, at 466). The Court went on to stress that "the wire tapping here disclosed did not amount to a search or seizure within the meaning of the Fourth Amendment"(Id.). In his visionary dissenting opinion, Justice Louis Brandies warned against the advances of technology that could help government carry out espionage against citizens. "The progress of science" he wrote, "in furnishing the government with means of espionage is not likely to stop with wire tapping. Ways may some day be developed by which the government, without removing papers from secret drawers, can reproduce them in court, and by which it will be enabled to expose to a jury the most intimate occurrence of the home" (Id. at 571). His opinion resonates the argument made some forty years earlier in his article, coauthored with Samuel Warren, in which they argued that common law recognized principles of privacy. Nonetheless, the near majority of the Court justices in this case did not recognize the constitutional right of individuals to information privacy.

The Olmstead decision was overturned in 1967 when the Supreme Court ruled in *Katz v. United States* that the use of electronic surveillance by the police, although exerted on a public telephone booth, was unconstitutional because "Fourth Amendment protects individuals, not places" (389 U.S. 347, 88 S.Ct. 507, U.S. Cal., 1967, at 349). Charles Katz was a professional gambler who used a public telephone to place his bets or to get the results. The FBI attached an electronic hearing and recording device to the exterior of the public telephone booth which Katz used every morning to place his bets. In this landmark decision, the Court asserted, "What a person seeks to preserve as private, even in an area accessible to the public, may be constitutionally protected under Fourth Amendment" (Id.). Nevertheless, the Court proceeded to warn that "the Fourth Amendment cannot be translated into a general 'right to privacy' " (Id. at 347). In his concurring opinion, Justice John Harlan articulated a two-fold requirement to determine the extent to which the Fourth Amendment affords protection against violations of privacy. Justice Harlan's formula requires that (1) the individual exhibit an actual (subjective) expectation of privacy; and (2) the expectation be recognized by the society as "reasonable." The Katz decision and this formula have since played a crucial role in many privacy cases, however, they do not seem to go beyond the Fourth Amendment intent to curb the government ability to conduct unreasonable searches and seizures, hence falling short of restricting malicious invasion of privacy by private parties.

In *United States v. Miller,* the Supreme Court decided that individuals had no expectations of privacy for the information supplied voluntarily for commercial use (e.g., banking services). Mitch Miller was charged with several federal offences, including tax evasion and the intent to defraud the government of tax revenues. The court subpoenaed Miller's bank records, which were subsequently provided by the bank. Miller filed a motion to suppress the bank materials on the grounds of the Fourth Amendment protection of privacy. The Supreme Court held that the subpoenaed documents were the bank's business records, not Miller's private papers, and therefore were not protected by the Fourth Amendment. It appears as if the individual forfeits the privacy of personal data once such information is provided for commercial purposes. However, this may well be a misreading of the Court's ruling, because the Court found "no legitimate 'expectation of privacy' in their [checks] contents. The checks are not confidential communications but negotiable instruments to be used in commercial transactions" (425 U.S. 435, 96 S.Ct. 1619, U.S. Ga., 1976, at 1624). Since a great deal of controversy surrounding information privacy pertains to commercial use, it remains an open question whether the commercial use of the Internet would fall under this decision and not be protected by the constitutional right to privacy as expressed in Katz: do online shoppers have a "reasonable expectation of privacy" when supplying their information for commercial use? The answer remains far from clear.

In 1977, another important case was presented before the Supreme Court. In *Whalen v. Roe*, a New York statute to centralize information about the users of dangerous prescription drugs was challenged. In an effort to prevent the abuse of dangerous drugs and their unlawful reselling in the market, the New York statute required that a copy of the prescription, along with the patient's and physician's pertinent information, be filed with the state's Department of Health and be input into a computerized databank. Patients and physicians filed a lawsuit for the state's violation of the constitutional protection of the doctor-patient relationship under the Fourteenth Amendment. The Supreme Court upheld the New York statute, which in its opinion had not violated any constitutional rights. The Court expressed confidence that the state had considered adequate security to safeguard the privacy and identity of patients. The Court, too, relied on the successful implementation of similar databanks in California and Illinois, and their immense help in the identification of violators. Yet, the Court recognized the threat posed to privacy by "the accumulation of vast amounts of personal information in computerized data banks or other massive government files" (429 U.S. 589 97, S.Ct. 869, U.S. NY., 1977, at 605). However, the Court stressed that such information gathering was a necessary part of public services, such as welfare, social security, unemployment, and health system. The Supreme Court proceeded to define a "zone of privacy," which in its opinion involved two separate interests: "One is the individual interest in avoiding disclosure of personal matters, and another is the interest in independence in making certain kinds of important decisions" (Id. at 599–600). The Supreme Court did not believe that the New York statute violated either interest. The Court's decision in *Whalen v. Roe* is an attempt to balance privacy interests against government's interest in implementing a sound public service.

Numerous privacy decisions made by the United States Supreme Court over more than a century demonstrate a marked metamorphosis of the concept of privacy. Paradoxically, the zone of privacy has expanded to include nearly all sorts of private information regardless of whether or not it is contained in a private space; yet the information technology exerts a greater impact on privacy while the society as a whole, too, seem to enjoy a greater access to personal information. The interplay of these factors has given rise to the emergence of the notion of "balance" between public and private interests. Though the Supreme Court's decisions have made outstanding contributions to the evolution of this notion, the answer to the ever-increasing concerns about privacy hardly seems to lie in the Constitution, because these concerns are no longer limited to the government's ability to intrude into citizen's private affairs. This has created the need for effective legislation. It may also have rendered the notion of "balance," which is a product of the policy debates of the 1960s, outdated.

Common Law Remedies of Privacy Violations

Privacy rights have long been established in the United States' common law. The early cases date back as early as 1881 in *DeMay v. Roberts* where, albeit without mentioning violation of privacy, tort relief was granted to the plaintiff for having been observed by the defendant during the childbirth. In *Pavesich v. New England Life Insurance Company* in 1905, the Georgia Supreme Court made extensive efforts to derive the right to privacy from natural law, thereby establishing a natural right to privacy.

In their groundbreaking article, Samuel Warren and Louis Brandeis argued for the existence of privacy rights in the common law. "The common law," they wrote, "secures to each individual the right of determining, ordinarily, to what extent his thoughts, sentiments, and emotions shall be communicated to others" ("The Right to Privacy," 4 *Harvard Law Review* 193 (1890); reprinted in *Philosophical Dimensions of Privacy: An Anthology* (1984), p. 78). This view posits the central role of "information" and "communication" in the privacy discourse. More important, Warren and Brandeis anticipated the impact of new technologies on privacy. "Instantaneous photographs and newspapers enterprise have invaded the sacred precincts of private and domestic life; and numerous mechanical devices threaten to make good the prediction that 'what is whispered in the closet shall be proclaimed from the housetops' " (Id. at 76). Interestingly, Warren and Brandeis recognized circumstances under which the individual's right to privacy must be balanced against other rights and/or interests, including public interests.

In a classic article published in 1960, William Prosser, in line with Warren's and Brandeis's argument, asserted that "the right to privacy, in one form or another, is declared to exist by the overwhelming majority of the American courts" ("Privacy [A Legal Analysis]." *California Law Review* 1960, 48:338-423; reprinted in *Philosophical Dimensions of Privacy: An Anthology* (1984), p. 106). He went on, however, to assert that privacy rights concern different interests, as opposed to an overarching right as implicated in Warren's and Brandeis's analysis. Enumerating a large number of privacy cases, Prosser set forth four distinct torts, with little underlying commonality, that may take place in violation of privacy. These four torts are "(1) Intrusion upon the plaintiff's seclusion or solitude, or into his private affairs; (2) Public disclosure of embarrassing private facts about the plaintiff; (3) Publicity which places the plaintiff in a false light in the public eye; and (4) Appropriation, for the defendant's advantage, of the plaintiff's name or likeness." (Id. at 107)

Despite different representations of privacy rights in Prosser's account, attempts can be made to apply the above torts to violations of information privacy. The first tort may take place when, for example, an Internet user, while surfing the Web or writing an email, is watched surreptitiously by an intruder (e.g., an operator, government agent, or a hacker). Likewise, private facts about one's life can be disclosed on the Internet or by sending mass emails with only little effort (second tort). The same mediums can be utilized to spread false information about an individual (third tort). No less important is the rising concern over identity theft and the abuse of PII stored in electronic format, either in one's computer hard disk drive, or during an online communication, or in a financial institution's database (fourth tort).

William Prosser's articulation of privacy interests was criticized in 1964 when Edward Bloustein contended that privacy right is not a composite of a number of interests, but rather an inherent social value, that is, the value of human dignity. Bloustein believed that Prosser had missed Warren's and Brandeis's point that the right to privacy could not be reduced to other rights or interests. Collection and centralization of citizens' information in government databanks, Bloustein held, were infringing upon citizens' privacy. He noted, however, that "the social benefit to be gained in these instances require the information to be given and that the ends to be achieved are worth the price of diminished privacy" ("Privacy as an Aspect of Human Dignity, An Answer to Dean Prosser." *New York University Law Review* 39: 962–1007, 1964. reprinted in *Philosophical Dimensions of Privacy: An Anthology* (1984), p. 185; emphasis added).

Despite their philosophical differences, these three influential theories converge at the point that common law offers a viable source to address the right to privacy and to settle its violations accordingly. In the 1971 Senate hearings conducted by Senator Sam Ervin, Jr., reference was made to the privacy protection afforded in the common law. Nonetheless, the question remains as to the extent to which common law remedies of privacy invasion are applicable to information privacy in the digital era. There is little doubt how fundamentally the widespread use of information technology has displaced the public and private boundaries. "Public" and "private" have a conventional character and social conventions undergo changes over time; so do standards of "expectations of privacy" and "reasonableness." The ever-changing nature of information technology calls for new interpretations of common law as to how the privacy of personal information in electronic format may be defined and how its violations may be remedied. This further underscores the necessity of establishing a new privacy framework and a baseline for privacy standards.

The United States Congress has enacted numerous statutes to dovetail the constitutional protection of privacy. These legislations have adopted a sectoral approach via safeguarding privacy within the context of socioeconomic sectors, hence creating a patchwork of privacy laws. The following is a snapshot of the most important privacy statutes along with a brief analysis of their effectiveness in protecting information privacy.

Statutory Protection of Privacy

Privacy Act (5 U.S.C. §552a; 1974, 1994, and Supplement 1996)

As discussed earlier, Congress enacted the 1974 Privacy Act, based on the HEW's 1973 report, to alleviate public worries about the implications of computerization for personal privacy and to limit government's ability to collect and disclose personal information maintained in federal agencies. The Act clearly observes threats posed by computer technology to personal privacy and, accordingly, sets

Principles of Fair Information Practices based on which government entities may collect and divulge personal data. The Act proscribes the disclosure of personal information without written consent from the individual who is the subject of the information. According to the Act, anyone about whom records are kept in federal agencies must be able to access and challenge the content of the pertinent information. In an attempt to balance personal privacy against the society's right to know, the Act provides exceptions as required under the Freedom of Information Act (5 U.S.C. §552).

The Privacy Act of 1974 is structured in such a way as to ease concerns about government's access to too much information on individuals without their knowledge, so it is not capable of addressing threats posed by corporate entities and other private parties who may use technology to acquire private information. The Act has furthermore been criticized for its failure to meet its goals, which for the most part has to do with the lack of effective scrutiny and of proper enforcement.

Right to Financial Privacy Act (RFPA; 12 U.S.C. §§ 3401 et seq., 1978)

Like the Privacy Act of 1974, the Right to Financial Privacy Act was enacted to curb the government's ability to access individuals' financial records maintained by financial institutions. The terms of the RFPA appears prima facie to run counter to the Supreme Court's decision in Miller where the Court asserted that individuals had no expectations of privacy once the personal information was voluntarily supplied for commercial purposes. The Act includes exceptions such as where the disclosure is authorized by the individual and where a search warrant or a judicial subpoena is provided. Additionally, in the event government intends to access financial records, the individual must be notified of the intent to search and be given a chance to challenge the access. As the RFPA, in keeping with the spirit of the Fourth Amendment, is designed to put a curb on government's ability to conduct unreasonable searches and seizures, it does not seem potent enough to limit financial institutions' ability to exchange their clients' information among themselves. Thus the Act cannot serve to address the increasing concerns due to expanding online shopping, the sale of consumers' data, and the practice of data profiling which has mushroomed in the business community.

Electronic Communication Privacy Act (ECPA; 18 U.S.C. §§ 2510–2521, 2701–2710, 3117, 3121–3126)

Comprised of three titles, the ECPA takes a broad approach to the protection of electronic information and communications against interception and surveillance. Title I protects information privacy by proscribing interception of various types of electronic communications. Although the Internet and email are not mentioned in the Act's definition of electronic communication, they seem to be covered by the direct reference to "any computer facilities or related electronic equipment for the electronic storage of such communications" (Ibid., 2510(a)(14)). This protection is effective insofar as the communication "affects interstate or foreign commerce" (Ibid., 2510(a)(12)), so the noncommercial communications seem to have been left out of the scope of the Act.

Title II concerns unauthorized access and disclosure of stored data without express consent of the subject of the data. Service providers are prohibited to disclose such data unless the disclosure is necessary for the regular maintenance or is required by law. Violators of this title are subject to punishment as set forth in the law "if the offence is committed for purposes of commercial advantage, malicious destruction or damage, or private commercial gain" (Ibid., 2701(b)(1)). Like Title I, Title II does not extend the privacy protection to noncommercial stored data. A government agency may access contents of stored communications once a warrant has been obtained. As mentioned at the outset of this chapter, Title II of the ECPA was cited in the DoubleClick Privacy Litigation, but the federal court held that Title II was meant to prevent hackers from accessing data, so it would not apply to a service provider whose access was a necessary part of fulfilling its service agreement.

Title III protects privacy of information generated by such electronic devices as pen registers and mobile tracking devices that are not specifically related to the subject matters discussed in this chapter.

Computer Fraud and Abuse Act (CFAA; 18 U.S.C. § 1030, 1994)

This Act makes it unlawful to use a computer without authorization, or by exceeding the authorization, to acquire "any restricted data," such as those "defined in paragraph y. of section 11 of the Atomic Energy Act of 1954" (Ibid., 1030(a)(1)). As the Act is intended to contain the abuse of computer technology in financial institutions and/or the U.S. government, it falls short of protecting computer devices used by private parties or private information obtained by unauthorized access to a personal computer.

Health Insurance Portability and Accountability Act (HIPAA, Pub. L. 104–191)

Enacted in 1996, the HIPAA's goal is to lower the cost of health care through adopting for sets of standards: (1) identifier standards; (2) transaction and code set standards; (3) privacy standards; and (4) security standards. Privacy standards are intended to protect personally identifiable health information. Pursuant to HIPAA (Ibid., Sec. 264(a)), privacy standards were proposed in October 1999 by the Secretary of Health and Human Services (HHS) and became law in December 2000. The law required compliance by April 14, 2003.

The privacy standards consist of five main principles.

- Consumer control: it includes the right to receive a copy of medical records.

- Accountability: it sets penalties for violations of health privacy.

- Public responsibility: it seeks to balance privacy against public interests, such as protecting public health, preventing epidemic diseases, and promoting medical research that benefits the pubic.

- Boundaries: it requires that a patient's medical records be used for healthcare purposes only, rather than unrelated purposes such as hiring and firing.

- Security: it requires organizations (e.g., hospitals, providers, health plans, etc.) to implement adequate security procedures to protect patients' privacy.

The HIPAA is the first public law to establish national privacy standards of health records. There are concerns, however, that the HHS is under pressure to relax some of the privacy provisions of the Act. A recent letter to the Secretary of HHS by 30 activist groups, including the Health Privacy Project (HPP) and EPIC, warned that the financial services industry is lobbying to gain access to critical medical records for data-mining purposes. The letter appealed to the Secretary not to give in to such anti-privacy pressure, which would constitute violation of the HIPAA.

Children Online Privacy Protection Act (COPPA; 15 U.S.C. §§ 6501 et seq., 1998)

The COPPA is specifically structured for the protection of children's online privacy. The Act requires an online service "to provide notice on the website of what information is collected from children by the operator, how the operator uses such information, and the operator's disclosure practices for such information; and to obtain verifiable parental consent for the collection, use, or disclosure of personal information from children" (Ibid., 6502(b)(1)(A)(i) and (ii)). The Act has also encouraged websites targeting minors to undergo self-regulation by following the provisions specified by the COPPA, hence creating a safe harbor. An eligible website would be required to comply with its guidelines that "after notice and comment, are approved by the Commission [FTC] upon making a determination that the guidelines meet the requirements of the regulations issued under section 6503 of this title" (Ibid., 6503(b)(2)). Pursuant to the COPPA, the FTC issued its

Rule for the enforcement of the Act's provisions. The FTC regulations promulgated under this Act took effect on April 21, 2000 (16 C.F.R. §312). Many children websites have since applied for safe harbor status. The COPPA, along with subsequent self-regulatory measures, has proved to secure a significant degree of privacy for children while holding the websites accountable under the Act.

Gramm-Leach-Bliley Financial Services Modernization Act (GLBA; Pub. L. 106–102, codified at 15 U.S.C. §§ 6801 et seq.)

Enacted in November 1999 to modernize financial services, the GLBA is one of the most important federal legislations that aimed to protect financial privacy. The Act makes it an obligation for financial institutions to "respect the privacy of customers and to protect the security and the confidentiality of those customers' non-public personal information" (Ibid., 6801(a)). Financial institutions are required by this public law to disclose their privacy policy when a customer seeks service and a relationship is being established. Pursuant to Title V of the Act, FTC has issued its Privacy Rule which sets the ground for the implementation of the privacy provisions of the Act (65 Fed. Reg. 33, 646; 16 C.F.R. §313). Privacy activists, such as U.S. PIRG, have complained that Title V does not provide adequate protection of privacy since it is based only on notice. That is to say, upon providing notice to their customers, financial institutions can share personal information with their affiliates and third parties, including telemarketers.

The USA PATRIOT Act of 2001 (Uniting and Strengthening America by Providing Appropriate Tools Required to Intercept and Obstruct Terrorism; Pub. L. No. 107–56, 115 Stat. 272)

Introduced in the wake of the September 11 terrorist attacks, this public law incorporates provisions of two earlier anti-terrorism bills (H.R. 2975 passed the House on October 21, 2001; and S. 1510, passed the Senate on October 11, 2001). The USA PATRIOT Act amended more than 15 legislations, including ECPA, FCRA, RFPA, and CFAA, in order to enhance government's authority to access personal information available in electronic or any other formats. Among other anti-terrorism procedures, the law amends the ECPA in such a way as to provide government with a greater authority to intercept suspected terrorist communications. Law enforcement and investigative agencies are given unprecedented latitude to discover terrorist plots. Any business that holds customers' personal data, especially Internet service providers who have greater access to individual's information and communications, will be required to cooperate with government agencies. The Act seems to have reversed legislations that over the past three decades have aimed at limiting the government's ability to surveil citizens. On the other hand, the Act has once again raised the issue of balancing privacy against other public interests. The anti-terrorism campaign and security concerns, however, seem to have tilted the balance away from privacy interests.

According to a sunset clause of the USA PATRIOT Act of 2001, certain amendments of the Act that enhances government's surveillance authority will terminate on December 31, 2005. The sunset clause does not apply to a number of amendments, including the extension of per register, trap and trace authority to the Internet. In November 2003, at the behest of the Bush Administration, a joint House-Senate Conference Committee approved a provision that will further expand government's authority under the USA PATRIOT Act. This will allow the FBI to obtain records, without a court warrant, from a number of businesses, including securities dealer, post office, car dealers, casinos, and travel agencies. At the same time, the Senate Judiciary Committee has begun hearings on the impact of the Act.

Other Federal Statutes

A large number of statutes have been enacted to protect privacy interests at different levels. Privacy of mail (39 U.S.C. § 3623, 1994) proscribes access to the content of mail by parties other than the addressee. The information associated with delivery of mail is exempted from this proscription. Wiretap statutes (18 U.S.C. §§ 2510 et seq., 47 U.S.C. § 605) have also aimed at safeguarding electronic communications and preventing unauthorized interception. Authorized interceptions under these statutes include, but are not limited to, regular maintenance of common carriers and employees' business-related communications in workplace; privacy of employees' communications that are not related to business has proved controversial. The Telecommunication Act of 1996 (Pub. L. No. 104–104, § 222, 110 Stat. 56, 1996) also sets rules for providers of telecommunications services to protect customers' personal information and to avoid its divulgence without express consent of the subject of data or as required by law. Computer Matching and Privacy Protection Act of 1988 (Pub. L. No. 100–503) regulates exchange of computerized records among government agencies and establishes a Data Integrity Board within each agency to oversee the sound implementation of the law.

The above-mentioned legislations, and many others that are not discussed here, are strictly targeted toward specific aspects of private information and therefore cannot afford broad interpretations to extend to other areas of information privacy. This sectoral approach is, to some extent, inevitable because it does not seem practical to cover diverse dimensions of privacy in a comprehensive law. These partial protections of privacy, however, are likely to neglect other important aspects of private information or to fail to keep up with the rapid evolution of technology. They are also likely to present inconsistency with one another. This is in part the reason why privacy activists have long advocated the establishment of a baseline privacy standards via federal legislation. The Privacy Act of 1974 is said to have aimed such a baseline in accordance with the Code of Fair Information Practices. Federal agencies would have been required to develop guidelines and regulations for different socioeconomic sectors. The legislators' failure to reach a consensus about this baseline, as well as the conflict between the stakeholders' interests, kept the Act from meeting its goal.

Self-regulatory Approach to Privacy

Since the advent of information technology, a number of government agencies have been involved in the protection of information, and especially online, privacy. This is an unintended result of the sectoral policy approach, which for the most part relies on the industry to undergo self-regulation. Pursuant to the Federal Trade Commission Act of 1914 (FTCA, 15 U.S.C. § 41 et seq.), the FTC was established by the Congress and was granted the mandate to prevent businesses from unfair competition or deceptive practices. Historically, the FTC has functioned as the government watchdog to establish the rules of fair business practice and, to the extent possible, to enforce them. To that end, the FTC has extended the widely accepted Principles of Fair Information Practices to online communication. Thus, given the nature of the FTC mandate, information and communication privacy seem to have over time transformed to a consumer issue, hence excluding non-commercial aspects of privacy.

In its 1998 report to Congress, the FTC called attention to the inadequacy of legal protection of online privacy. "Current American privacy law," FTC reported, "can best be described as sectoral consisting of a handful of disparate statutes directed at specific industries that collect personal data and none of which specifically covers the collection of personal information online" (Privacy Online: A report to Congress. June 1998. p. 62, fn. 160). Nonetheless, the priority has been placed on the adoption of a self-regulatory regime, which is believed "to respond quickly to technological changes and employ new technologies to protect consumer privacy" (Id. at 41)

The FTC has put forward guidelines for self-regulation and has encouraged the online industry to adopt the five core Principles of Fair Information Practices in the collection and exchange of personal data: (1) notice/awareness; (2) choice/consent; (3) access/participation; (4) integrity/security; and (5) enforcement/redress. Self-regulation, according to the FTC report, requires, "at a minimum, institutional mechanisms to ensure that consumers have a simple and effective way to have their concerns addressed…If the self-regulatory code has been breached, consumers should have a remedy for the violation" (Id. at 11). During the FTC workshops, the online industry expressed interests in self-regulation, however, "very few provide any kind of enforcement mechanism, an essential element of effective self-regulation" (Id. at 15). To measure the effectiveness of self-regulatory policy, in June 1997, the FTC conducted a survey of 1402 commercial websites. Among other findings, the survey found that while 92 percent of the websites collected personal information, only 14 percent of them posted any notice or disclosed their privacy policy (Ibid., pp. 19–27). Nevertheless, the FTC continued its self-regulatory programs while at the same time encouraging the online industry to create their own programs. There has been a significant increase in recent years in the number of commercial websites that adhere to the Principles of Fair Information Practices and several self-regulatory programs, such as Seal Programs, TRUSTe, BBBOn-line, AICPA, and CPA WebTrust, have been initiated by the industry.

Self-regulation has proved a challenging undertaking for both the FTC and the online industry. The majority of web-based businesses post their privacy policies and provide an option to opt out. In its 2000 report to Congress, FTC recognized inefficacies of self-regulation and, accordingly, recommended that federal legisla-tion be enacted. The FTC observed that, for example, while most sites (57 percent of random sample and 78 percent of most popular group, as defined by the survey) that had adopted privacy protection measures placed third party cookies on visi-tors' computes to track their surfing and purchasing behavior, a small fraction (22 percent of random sample and 51 percent of most popular group) informed visitors of such practice (*Privacy Online: Fair Information Practices in the Elec-tronic Marketplace*; A Report to Congress. May 2000. p. 21).

Following the 2000 presidential election, the FTC witnessed a change in its chairmanship, which consequently affected its recommendation to the Congress. While committing to a pro-privacy agenda, the FTC questioned the need for a broad-based privacy legislation. In his remarks at the Privacy 2001 Conference on October 4, 2001 in Cleveland, Ohio, FTC Chairman, Timothy J. Muris pointed out that, due to inherent difficulties involved in legislating a federal privacy statute, the self-regulatory initiatives needed more time to bear satisfactory results. The decision to enact a federal law, he further argued, must be backed up by a cost-benefit analysis.

Prospective Solutions to the Privacy of Electronic Information and Communication

Legal Protection of Privacy

There is little doubt about the crucial role of laws in protection of privacy. Absent legal obligations, few government agencies and private organizations would will-ingly incur a variety of costs involved in the implementation of protective measures. Despite the partial success of the self-regulatory policy, the "invisible hand" of the market will likely exert little effect in this area.

The much-debated issue is the applicability of old privacy laws that have been amended for the new information and communication technologies. It remains dubious how effectively such laws may protect privacy rights in the context of information and communications technologies. Further, the enforceability of privacy laws, their execution, and official oversight are among the main requisites of successful legislations. Advocacy groups have criticized the lack of sufficient enforcement for existing laws. The lack of effective enforcement is in part due to the absence of an oversight agency. The FTC's oversight role has been dominated by its self-regulatory policy, which calls for adequate time to bear fruits. After all, the FTC's oversight is limited to commercial aspects of privacy. Moreover, the vast majority of the existing statutes, however effective they may be, protect personal privacy against government intrusion as well as against corporate abuse of consumers' data; individuals' violation of privacy is left to the common law whose applicability and effectiveness in the context of information and communication technologies is in question.

Privacy activists have long argued that privacy laws are geared toward data protection, which has in effect made them information laws, while little emphasis has been placed on surveillance and other non-informational aspects of privacy. Privacy laws, they further argue, also allow violations on such grounds as national security and law enforcement, and do not prohibit excessive collection of data, but at best protect existing data. Whereas the FTC has remained reluctant to recommend any comprehensive legislation, a number of legislation proposals has been introduced in recent years at the House and the Senate, and testimonies by privacy advocates, scholars, victims of privacy violations, and industry representatives have shed further light on the complexities of privacy matters. Yet, there exists a sharp schism as to whether or not an omnibus legislation proffers the solution to privacy.

Legislation and self-regulation have been perceived as mutually exclusive. This, however, may have misled policymakers and experts. Absent a general policy framework structured by statutes, self-regulation is unlikely to meet its goals. As a practical matter, it is hard to fashion laws that can cover all sectors, but the ineffectuality of the current legal protections needs to be mended anyway. Thus a new legal regime—though not necessarily a European style omnibus—can function as a deterrent of privacy invasion at different levels and with regards to different socioeconomic sectors while at the same time stimulating both self-regulation and the manufacture of PETs. It can also outlaw the use of privacy-invading technologies (PITs) by private parties and can include recoveries for victims of privacy violations. Such legislations must enunciate new privacy principles that take into account the widespread use of information technology in everyday life. While creating a policy framework, a new federal legislation could establish an oversight agency responsible for the enforcement of privacy laws and for making recommendation to Congress for new laws as necessary. The legislation can revisit the efficacy of conventional methods of privacy protection, such as opt-out option, which provides limited choice to users, since by default it allows providers of online and offline services to collect personal information, or share it with their affiliates and third parties, without having to ask users for their consent. Moreover, many people are not aware of such a choice, or simply do not have time and patience to read privacy policies that are permeated with legal jargons and loop-holes. The legislation also needs to include a sunset clause to provide for periodical (e.g., every five years) evaluation of the law and for making recommendations when necessitated by the change of technology and/or by socioeconomic conditions.

No less important is the desired effects of such legislation in promoting trade between the United States and Europe. In 1995, the European Union (EU) issued its first directive on privacy to create a Safe Harbor Program. The directive prohibits the transmission of personal data to non-EU nations that do not meet the European adequacy standards for data protection. The directive failed to define adequacy, however. In collaboration with the European Commission, the U.S. Department of Commerce has developed a self-regulatory safe harbor program in order to assist American businesses to engage in commercial relationship with their European counterparts without facing legal charges due to the inadequacy of their privacy policies. This program was approved by the EU in July 2000 and has since become an important element in Trans-Atlantic trade; it provides assurance to European companies that the American companies certified by this program meet a minimum level of privacy standards.

The EU Safe Harbor Program requires compliance in seven areas, which include notice; choice; onward transfer (or transfer to third parties); access; secu-rity; data integrity; and enforcement. These are very similar to the U.S. Code of Fair Information Practices as well as to FTC's self-regulatory guidelines. As mentioned earlier, however, the FTC dropped "enforcement" from the list of fair information practices in its 1998 report to Congress and vaguely addressed "onward transfer." The global trade opportunities can create incentives for Amer-ican online businesses to undergo self-regulation, but effective legislation can redefine a baseline of privacy standards, including those that are common with the EU directive, while giving the self-regulatory regime more teeth.

Technological Solution

If information and communications technologies have contributed to the erosion of privacy, similar technologies are the would-be antidotes. From a technological standpoint, privacy is a technical ability to control access to, or the release of, information about oneself; such ability is taken into account in design of PETs. A variety of PETs has emerged over the past decade and they vary from encryption techniques to blind signatures. Security measures are by far the most important technological prerequisite of privacy protection (see Chapter 5 in the present volume for a detailed discussion of security issues).

PETs zero in on concealing the identity of the subject of data. One such technology is a smart card with 32 kilo bytes of memory that stores up to 7 different types of information, including personal identity, driver's license, bank accounts, healthcare, and immigration status (*MIT Technology Review*, May 2003). The Malaysian government is currently in the process of using this new technology throughout the country. The smart card makes the pertinent information available only to authorized entities concerned. For example, only a bank teller will be able to access the account information while a police officer is the only one who can access the driver's license information. Such technologies empower individuals to gain control over their personal information by choosing the extent to which particular information may or may not be disclosed.

P3P, developed by the World Wide Web Consortium (W3C), is another example of technological solution that enables the users to set their privacy preferences and to identify privacy-friendly policies by exchanging codes between computers and websites. This, however, requires that a website adopt P3P standards and protocols, and configure its privacy policy in a machine-readable format. As mentioned earlier, in their report of June 2000, EPIC and Junkbusters took on a sharp criticism of P3P for being "a complex and confusing protocol that will make it more difficult for Internet users to protect their privacy" as well as for its inability to minimize or eliminate the collection of PII (*Pretty Poor Privacy: An Assessment of P3P and Internet Privacy*, p. 1).

PETs are unlikely to flourish unless there exists a socioeconomic demand and also legal requirements are set by the law. It appears that the social demand (i.e., awareness about the threats to privacy and the necessity of its protection) is on the rise, so privacy-enhancing features in new technologies will become a must-have for more and more users. Such a social demand has created its own market where the manufacturers are competing to build superior privacy-enhancing features into their new products. This competition is expected to drive down the cost of PETs over time. New legislations, too, can reinforce the importance of privacy protection in design and manufacture of new technologies. Drawing an analogy to the automobile design and manufacture is instructive: the public is aware of the necessity of safety features, such as air bag and seatbelt; manufacturers compete to produce automobiles with such features; and it is required by law to ensure that safety features are built into new automobiles.

Ironically, there is also a demand for PITs regardless of how legitimate their use may be (e.g., law enforcement agencies, parents protecting their children from child predators). TrueActive is one such manufacturer who, conscious of public concerns, potential liabilities of the use of snooping features and ethical burdens, chose to remove from its software "silent deploy"—a feature that makes it possible to install the snooping software on a target computer via email ("Snooping Software Gains Power and Raises Privacy Concerns," *The New York Times*, October 10, 2003).

There are software products, such as iSpyNow and LoverSpy, that render online snooping possible. It can be envisioned that the demand for such software products, no different from the demand for illegal drugs and weapons, will continue to exist. But, at a minimum, the government initiative to outlaw such software products will minimize the risk of privacy invasion.

Conclusion: An Amalgam of Solutions

Information and communications technologies have made enormous contributions to the faster growth of the economy by raising efficiency and productivity in economic sectors while introducing new sectors, like the online industry, which has enabled Internet users to benefit from a wider array of services. The average American, however, seems unwilling to forgo her/his privacy in return for new services and conveniences. Nevertheless, part of the costs associated with the rendition of computerized services and the widespread use of the Internet has inevitably been paid by a shrinking "zone of privacy." Every day new technologies enter the market and soon find their way to public and private organizations; they are also widely used by individuals for personal purposes.

Information and communications technologies have metamorphosed the classical notion of privacy (i.e., "the right to be let alone"), thus calling for a new conceptual/legal framework to redefine and protect privacy. To be sure, information privacy issues are not solely caused by the new technology, but are created by the interplay of social, economic, political, legal, and technological forces. Thus, any solution to privacy problems must take into account the interaction of these forces. To be sure, a one-dimensional approach will be highly unlikely to capture the depth and complexity of the privacy issues. In spite of the urgent need for more efficacious laws, there is no guarantee that a comprehensive legislation alone could afford adequate protection of information privacy while keeping up with the constant changes of technology. What is needed is a unified policy framework to conjure up privacy principles that can accommodate an amalgam of solutions. These privacy principles would determine the desired balance between public and private interests, and how such balance can be accentuated in privacy laws and be built into new technologies.

Further Reading

Alan F. Westin (Project Director). *Databanks in a Free Society, Computers, Record-Keeping and Privacy.* New York: Russell Sage Foundation and the National Academy of Sciences, 1972.

Department of Health, Education, and Welfare. *Records, Computers, and the Rights of Citizens*. Washington, D.C.: Department of Health, Education, and Welfare, 1973.

Ferdinand David Schoeman (ed.). *Philosophical Dimensions of Privacy: An Anthology*. New York: Cambridge University Press, 1984.

Priscilla M. Regan. *Legislating Privacy: Technology, Social Values, and Public Policy*. Chapel Hill: University of North Carolina Press, 1995.

Philip E. Agre and Marc Rotenberg (editors). *Technology and Privacy: The New Landscape*. Cambridge, MA: MIT Press, 1997.

Amitai Etzioni. *The Limits of Privacy*. New York: Basic Books, 1999.

Chapter 7

Intellectual Property

Lorraine Woellert

"I solved the problem by owning my own copyright, so nobody can screw around with my stuff. Nobody can take Star Wars *and make Yoda walk, because I own it."*

–George Lucas (Kevin Kelly and Paula Parisi. "Beyond Star Wars" *Wired* 5(2), Feb 1997.)

In the late 1990s, *Star Wars* producer George Lucas pioneered the use of digital technology for the big screen, producing one of the first ever, big-budget movie blockbusters created in large part on digital video and computers. When Lucas needed an army for the first installation in his *Star Wars* prequel trilogy, he assembled one on his desktop instead of hiring thousands of extras. When he needed bizarre-looking aliens to populate a scene, he didn't call makeup; he designed them digitally. Entire action scenes were "filmed" in virtual reality. And when the movie was edited, most of the work was done on a computer; the cutting-room floor was proverbial. In a 1997 interview with *Wired* magazine, Lucas predicted that it wouldn't be long before a couple of kids in a garage could produce a high-quality movie using relatively cheap digital equipment.

Lucas soon would discover, to his dismay, that he was right. Digital technology had helped Lucas re-invent filmmaking, but it also helped a legion of fans reinvent his artistic vision. When *Star Wars I: The Phantom Menace* hit the big screen in 2001, fans and critics were hugely disappointed. One reason: a floppy-eared alien named Jar Jar Binks. Jar Jar was a virtual character, created on a computer and digitally inserted into the plot. Jar Jar spoke his lines perfectly and interacted with his surroundings and the human cast. He was a nifty trick, but moviegoers weren't impressed. Jar Jar was stupid, fans complained. He slowed the plot, critics said. Even academics weighed in, accusing the wholly digital Jar Jar of perpetuating racist stereotypes.

It wasn't long before the tech wizardry that gave birth to Jar Jar brought about his demise. Mike J. Nichols, an avid *Star Wars* fan living in Santa Clarita, California, thought he could improve Lucas's film. Nichols loaded the *Phantom Menace* onto his PowerMac, where he used widely available software to digitally edit the movie and all-but-eliminate Jar Jar Binks. Nichols then posted his work—the *Phantom Edit*—on the Internet. *Star Wars* fans raved. They copied the rogue edit repeatedly and passed it around on line. Nichols's derivative work won a big following. The unofficial *Phantom* even worked its way into the mainstream, earning a favorable review in the *Chicago Tribune*. Before long, bootlegged videotapes were being sold on street corners. And subsequent "fan edits" began making the rounds, some of them dubbing Jar Jar's lines with a subtitled alien language that allowed revisionists to rewrite the film's dialogue to make the irritating foil quiet and wise.

For Lucas, imitation was not the sincerest form of flattery. After some initial bemusement, LucasFilm ordered Nichols to cease distributing copies of the *Phantom Edit* and accused Nichols of violating Lucas's copyright. Nichols complied, posting an apology to Lucas on his website and denouncing pirates who were profiting from the *Phantom Edit*. But the genie was out of the bottle, and it had a tough lesson for Lucas: In this digital age, laws protecting intellectual property simply can't cope with the technological realities that allowed *The Phantom Menace*—or any other digital creation—to be copied or altered, and distributed to a worldwide audience.

Digital Property: Promise and Perils

In the mid-1980s, creative works that once existed only in the tangible world of art, books, photographs, and phonographs began migrating to digital media. Copyright protections until recently were based on the premise that private copying would be laborious, time-consuming, and expensive. But digital technology changed all that, and the Internet changed the economics of distribution. Digital technology promised a world of new potential, offering clarity and malleability that analog could never deliver. But it also opened the door to illegal derivative works and widespread copies that have all the fidelity of the original, regardless of whether the copy is the first or the millionth. Movies can be uploaded from digital versatile discs (DVDs) onto any hard drive and watched anywhere on any computer. Songs are "ripped" from compact discs and posted on swapping websites, where they're available for free to millions of listeners who download them illegally onto blank CDs or convert them for play as MP3 files. Photographs have morphed from paper prints in an album to digital pictures on a computer, where they can be copied with ease and manipulated to change their content. A single purchase of software can be installed over and over onto any number of hard drives or copied onto floppies and passed around. And any digital content easily can be modified, or mutilated, using basic home equipment. Digital content has limitless plasticity: an image can be turned into a sound and vice versa, raising questions of authorship. Even quotidian web surfers are potential copyright infringers: the mere act of surfing the Internet requires a copy of each visited website to be cached on a hard drive, making copying a prerequisite of use.

Thus the digital revolution is an artists' dream and nightmare. Before content was digitized, it was easy to argue for the protection of creators. But content-protection heavyweights now are bumping up against new and compelling philosophical counterpoints, such as the need to foster technological innovation, which requires the freedom to reverse program, and the demands of tech-savvy consumers who are loathe to accept limits on how they can use their legally purchased material. The tension has led to a historical irony. Even as technology offers an explosion of easily accessed information, business, government, and the courts are moving to restrict that access.

The situation has created an expensive quandary for corporations from Microsoft to MGM, which have embraced technology's promise with new and better products even as those very products are reproduced illegally en masse in overseas factories, distributed for free online to millions of users, or simply handed off between friends. Content owners, from aging heavy-metal rockers Metallica to billionaire Bill Gates, are locked in a battle against consumer electronics manufacturers, influential cyber-libertarians such as Stanford's Larry Lessig, and legions of consumers for whom the digital revolution promised a fantastic new world where ideas could be given free reign and innovation could grow unfettered by old-fashioned notions of ownership.

Policy makers are caught in the middle. They have been called upon to weigh the needs of software programmers, Hollywood, the music industry and others whose livelihoods depend on their ability to market their creative works, against cultural and longstanding philosophical beliefs embedded in the Constitution, which recognizes democracy's reliance on a free exchange of ideas and information. At stake, too, is technological progress itself. The consumer electronics industry, led by companies such as Intel and Sony, argue that a heavy regulatory hand could throw cold water on the rollout of new computer technology, music players, games, and the like. All sides have had to contend with a seismic societal shift fueled by the freewheeling and youth-oriented Internet, which is redefining cultural norms and notions of ownership.

With economic, social, and legal institutions in play, striking an acceptable new balance between copyright law and fair use has been a tortuous, and so far fruitless, endeavor.

From the Constitution to Sonny Bono: A Brief History of Intellectual Property

Believing that a democracy could function only with an informed public, Thomas Jefferson fought aggressively to encourage the development of ideas and the free exchange of knowledge, which he believed should remain unfettered by ownership. In a 1813 letter to Isaac McPherson, Jefferson laid out the thinking behind his philosophy:

> If nature has made any one thing less susceptible than all others of exclusive property, it is the action of the thinking power called an idea, which an individual may exclusively possess as long as he keeps it to himself; but the moment it is divulged, it forces itself into the possession of every one, and the receiver cannot dispossess himself of it. Its peculiar character, too, is that no one possesses the less, because every other possesses the whole of it. He who receives an idea from me, receives instruction himself without lessening mine; as he who lights his taper at mine, receives light without darkening me.[1]

James Madison saw things differently. He sought to temper Jefferson's desire for unbridled intellectual exchange with an incentive for creators to continue creating. In other words, he wanted to give them the opportunity to profit from their works. After prolonged debate between these two giants of ideas, Madison prevailed. Intellectual property protection was codified in the U.S. Constitution, with Jefferson winning a key concession: copyright protections were limited to a finite time period.

---

[1]   Letter of August 13, 1813, to Isaac McPherson, "NO PATENTS ON IDEAS" accessed at *http://etext.virginia.edu/jefferson/texts/* (April 17, 2002).

Copyright's berth in the Constitution is testament to the founders' deep belief that a free exchange of ideas was vital to the workings of the new Republic. The Constitution sought to balance the free flow of ideas while encouraging the growth of knowledge by granting the U.S. Congress the power "to promote the progress of science and useful arts by securing for limited times to authors and inventors the exclusive right to their respective writings and discoveries."

The definition of "limited times" has been the subject of debate from the moment the phrase was put on paper. When Congress passed the Copyright Act in 1790, its protection extended 14 years and applied only to books, maps, and charts. But during the last two centuries, Congress has reacted to advancing technology, from piano rolls to computer software, by dramatically extending the rights of creators—corporate copyright holders in particular—and curbing public use (see Box 1).

Box 1: Copyright Act History

1790:	Congress passes the original Copyright Act.
1831:	Act expanded to include published music, extend protection to 28 years.
1856:	Copyright extended to cover plays.
1870:	Congress applies copyright to works of art. Lawmakers designate the Library of Congress as the central clearinghouse for copyright registration.
1897:	Public performances of music protected.
1909:	In the first extension of the act to technological reproductions, Congress protects the rights of musical compositions converted to player piano rolls.
1912:	Motion pictures protected.
1976:	Protection given to sound recordings published after 1972. Protection given to unpublished works.
1980:	Computer programs win copyright protection.
1992:	In one of the earliest battles over digital recording, the Audio Home Recording Act allows for hope taping of music and film for private use. To make the law amenable to copyright owners, the act levies a licensing fee on digital audio recorders; the technology never catches on.
1998:	The Copyright Term Extension Act, also known as the Sonny Bono Act, extends protection of corporate copyrighted works such as movies from 70 years to 95 years after the death of the artist or after the copyright was initially secured. The law applies retroactively, giving works scheduled to enter the public domain in 1999 a reprieve until 2019. In 2003, the Supreme Court declares the law constitutional, stating that Congress is within its rights to define what is "finite."

Congress won't be finished any time soon. Since recordings came under copyright protection in 1976, lawmakers have introduced more than 400 bills to change copyright law. In two centuries of debate over intellectual property, the key issue of how to balance the needs of the republic against the needs of the creator has never changed. In the process, two defining legal and cultural axioms emerged—fair use and first sale.

Fair Use

Fair use limits a copyright holder's monopoly on his work by allowing it to be used for purposes that serve the populace as a whole, such as education, news gathering, and satire. The Copyright Act of 1976, while giving copyright owners exclusive right to reproduce, distribute, perform, display, rent, or lend their creations, included exemptions to promote free speech, learning, scholarly research, and open discussion. Fair use assumes that such uses constitute a compelling enough social good that even if a copyright holder wanted to prevent such uses of their material, the law would not support them. Fair use has allowed a student to copy pages from a book, permitted political commentators to freely borrow and manipulate copyrighted images, and given newspapers the right to publish excerpts of copyrighted text and images for the purpose of informing the public.

In the 1970s, widespread use of new technology such as copy machines and videocassette recorders threatened to chip away at the fair use doctrine. A landmark legal battle pitting fair use against a new technology, analog Betamax video, augured the furor over digital content that would erupt 20 years later. In 1976, in a case that would become *Sony Corp. v. Universal City Studios Inc.*, television and movie studios sued the makers of Betamax videocassette recorders for copyright infringement. Their argument: videocassette owners who used Betamax machines to record television shows at home to watch later infringed on programmers' copyright. Studio executives pointed the finger at Sony and other manufacturers that marketed video recorders to the public, accusing the companies of contributory infringement. By selling technology that enabled people to record a broadcast—that is, make a copy of it—Sony was responsible in part for the copyright violation. A split Supreme Court disagreed. In a 5-4 decision the court in 1984 declared that "time shifting" a broadcast for personal use was a non-infringing, fair use of copyrighted material.

The court's deep divisions were testament to the critical cultural and economic issues at stake. The opinion was issued only after an unusually prolonged consideration that required two oral arguments. In the course of reaching a decision, the justices swung back and forth; the first vote after oral arguments was in favor of ruling against Sony. It took the court a year to render an opinion. In the end, the majority, led by Justice Harry Blackmun, an appointee of Republican President Richard Nixon, recognized society's need to foster new technologies, noting that copyright has "never accorded the copyright owner complete control over all possible uses of his work."

The majority also found that to prove infringement, copyright owners had to demonstrate damage to their market potential. But the use of videocassette recorders, Blackmun wrote, stood instead to increase broadcasters' viewership. The opinion proved prescient; instead of killing the movie and TV industry, video-cassette recorders spawned a lucrative secondary market for major films and shows, and established a primary market for low-budget productions.

First Sale

The first sale doctrine originally applied to books. It restrains copyright protection by limiting control over a particular copy of a work only until its "first sale"—the first time that a particular copy is sold. First sale dictates, for example, that the original purchaser of a work, be it a book, painting, or music recording, can then do what she wants with that copy—lend it, share it, resell it, manipulate its content, or destroy it. The doctrine has sweeping social implications. It is the basis for library lending, used-book stores, auction houses, and exchanges of books, music and other legally purchased property between friends and family. In the digital age, the first sale doctrine also has been applied to videotapes and DVDs, spawning a massive retail industry for movie rentals.

But the widespread application of digital technology ushered in new notions of ownership, beginning with the advent of mass-market software. When personal computers exploded onto the consumer market, software soon followed. Software developers had traditionally licensed, rather than sold, their programs to large commercial users. But it was impractical to bind millions of individual users to the same complex licensing terms when selling software off the shelf in retail outlets. In the mid-1980s, software developers began adopting what became known as a "shrink-wrap license." The shrink-wrap license, in theory, is a contract that sets terms under which the software developer sells the rights to use the product. The agreement, which is effective the moment the sealed packaging on a product is broken, limits the licensor's warranty on the product and the licensee's ability to use the product. The shrink-wrap concept soon moved on line, where it is known as a "click-wrap license." Before downloading software, a user first must agree to limited terms of its use, usually by clicking the "I Accept" button at the bottom of a long disclaimer. Shrink-wrap licenses have fundamentally challenged the first-sale doctrine's applicability to digital works.

Digital Revolution: A Shot across the Bow

The statistics were frightening: A federal survey found that 40 percent of Americans over the age of 10 had copied music in the past year. More than 1 billion recordings had been made at home. Worst of all, those surveyed believed that copying a song for personal use was acceptable. Doomsayers predicted nothing less than the collapse of the music industry.

But the year was 1988, years before the Internet became a household word and a decade before Napster's debut. The issue was home tape-recording. Digital content had just come into its own via widespread sales of music CDs, and a new Japanese technology, digital audio tape (DAT), was making its way to the United States. DAT was for tape recording what CDs were to vinyl records. Unlike the popular analog tape recorders in widespread use at the time, DAT could make absolutely perfect home recordings of a CD, and it offered an equally long lifespan. Recording studios, fearing a wholesale rip-off of their works, saw disaster looming.

The Recording Industry Association of America (RIAA), the music industry's powerful lobbying group, quickly pulled out all the stops to prevent DAT from making inroads to the U.S. consumer market. The RIAA pushed Congress to ban the sale of digital tape and mounted a heavily funded public relations campaign. The Consumer Electronics Association, led by Sony, fought back. It established a front group, the Home Recording Rights Coalition, which dispatched lobbyists to local record stores across the country to whip up grass-roots opposition to a DAT ban. Economists fanned out to survey teenagers about their music-taping habits in an effort to gauge home taping's impact on industry profits and artist royalties. The congressional Office of Technology Assessment opined on digital's challenges to copyright law, noting that as new technologies extended traditional bounds of copyright, "the major question facing Congress is whether extending copyright proprietors' rights to private use is necessary to serve the public interest."

Congress couldn't decide, so it dodged the question. In 1992, on a voice vote that relieved individual lawmakers of any accountability, Congress passed the Audio Home Recording Act, which imposed a 3 percent "piracy tax" on the importation, manufacture, and distribution of blank digital tape and digital recorders. The fee would be used to compensate popular artists such as Madonna and Michael Jackson, who presumably would see a decline in sales and royalties due to home copying. In an attempt at balance, the act protected music fans from prosecution for copying CDs for personal use. But that provision proved mostly moot because the legislation also required digital recordings to conform to the Serial Copy Management System (SCMS), an early digital rights management tool that used technology to limit reproductions. Home copying was allowed, but only once.

The music industry had won, but that victory would be short-lived. The DAT uproar was an early volley in the digital content wars that would erupt in full force less than a decade later. Thanks to the music industry's efforts, DAT never made huge inroads to the consumer market. But industry couldn't stand in the way of progress for long. The next generation of digital recording technology, coupled with cheap home computers that could make copies in the blink of an eye, prompted even more dire predictions about the collapse of the music industry. This time, the doomsdayers were closer to the mark; an upheaval was in the works.

WIPO and the WTO: Giving Birth to the DMCA

The 1995 Uruguay Round of trade talks, which led to formation of the World Trade Organization (WTO), sought to curb digital piracy by requiring its 150 member nations to provide standardized copyright protection to works from all other WTO-member nations. Under the agreement, the United States agreed to protect works that remained under copyright overseas but which might have fallen into the public domain in the United States because they did not meet requirements under domestic law.

A year later, the United States also signed on to two agreements drawn up by the World Intellectual Property Organization (WIPO), a group of 179 countries tasked with managing and protecting intellectual property rights. WIPO's roots went all the way back to 1873, to a meeting of the International Exhibition of Inventions in Vienna. Several inventors declined to attend that convention out of fear that their works would be copied and used in other countries. The Paris Convention for the Protection of Industrial Property was formed a decade later to find a way to offer global protection to ideas and inventions. The convention eventually became WIPO, which now is the intellectual property arm of the World Trade Organization. The WIPO Copyright Treaty and Performances and Phonograms Treaty of 1996 required its member signatories to protect copyrighted works from other member nations.

The treaties forced the United States to apply the same level of protection to non-U.S. works as was applied to U.S. works. At the insistence of U.S. negotiators, a key WIPO provision pushed member nations to pass laws that specifically protected digital intellectual property. The agreement also required nations to outlaw "circumvention" of any technology, such as encryption, that was used to protect digital copyrighted works such as CDs and DVDs. This provision on circumvention forced U.S. lawmakers to revisit the nation's 200-year-old standards of fair use and first sale.

The Congress, with both chambers dominated by Republicans, passed final legislation in October 1998. President Bill Clinton, a Democrat whose administration had taken a leading role in negotiating the WIPO treaties and the WTO agreement, signed the Digital Millennium Copyright Act into law on October 28, 1998.

The Digital Millennium Copyright Act

The Digital Millennium Copyright Act (DMCA) was the first comprehensive attempt by Congress to sort out the ownership and exclusive rights issues raised by digital technology. But instead of calming the debate, the 1998 act upset the precarious, longstanding balance that had existed between copyright owners and consumers. It became a lightning rod for controversy, pitting free-speech advocates, academics, artists, and consumers against corporate copyright holders from Hollywood to Orlando. Years after its passage, it remains a law in flux as multiple parties battle over its provisions in court, at the U.S. Copyright Office, in Congress, and in the press.

The DMCA's central provision—Section 1201—rewrote U.S. law to comply with the WIPO treaty. Section 1201 made it a crime to crack digital copyright protection measures such as encryption. It also made it illegal to "traffic" in such circumvention measures or post them on the Internet. While it made exceptions for academic research and engineering, those provisions remain vague. Even with the exemptions, the measure redefined copyright infringement and threw in to question long-accepted notions of fair use and first sale. For the first time in U.S. history, copyright infringers faced criminal fines and jail time.

The law had an immediate chilling effect on activity ranging from police investigations to academic research. In 1999, the police department in East Lansing, Michigan, was threatened with a lawsuit under DMCA for posting photos of wanted criminals on its website. The photos drew 60,000 viewers and led to 500 tips and 46 arraignments. But a freelance photographer claimed the police department violated the DMCA because it had posted his photographs on the Internet without his permission. The incident was a sign of things to come. Litigation over the bill began almost immediately.

Universal Studios v. Eric Corley

Section 1201 saw its first big court test in 1999, when eight major motion picture studios sued online publisher Eric Corley. Corley produced a webzine called *2600: The Hacker Quarterly*. One of his readers, a Norwegian teenager, posted several lines of code that could circumvent the Content Scramble System (CSS) that DVD manufacturers were using to prevent consumers from copying digital home movies. The Norwegian, Jon Johansen, alerted Corley to the code, called DeCSS, and Corley posted a link to it on his website, 2600.com.

DeCSS was no run-of-the-mill hack; it had a utilitarian purpose. The movie industry had designed CSS to allow Windows users to play DVDs on their home computers. The industry was working on a Macintosh version. But Linux users were left in the lurch. If users of the Linux operating system, which was growing in popularity, wanted to play DVDs on their computers, they were out of luck. A loose-knit hacking group called Drink or Die came up with a solution. The group's members reverse engineered CSS to write DeCSS in September 1999. Two months later the Linux utility was in widespread circulation on the Internet.

In the spirit of the open-source code movement, which promotes the free exchange of programming language (see section on Open Source Code), Corley publicized DeCSS so Linux users could decrypt their legally purchased DVDs for play on their computers. DeCSS also allowed home computer users to download DVD content to be played from a hard drive.

But by circumventing CSS, Johansen had violated the DMCA, Hollywood claimed. And by distributing that illegal code, Corely had "trafficked" in a encryption circumvention. The Motion Picture Association of America called DeCSS "akin to a tool that breaks the lock on your house." The fact that a lock can be picked is no excuse for burglary, industry argued. Reverse engineering was permitted under the DMCA, but only if it was used to ensure compatibility or interoperability with other programs. It was not permitted "if there was a readily available commercial alternative for that purpose," the MPAA claimed. "In this case, there exist MANY commercially available DVD players."

The motion picture industry also took aim at fair use, taking the fine point that the doctrine conferred the right to use something that was already available, not the right to access a work for fair use purposes. The DMCA, Hollywood argued, trumped any old-fashioned notions of fair use.

Stanford Law School Dean Kathleen Sullivan, one of Corley's attorneys, countered that programming code was a form of free speech and that restricting access to a software program could deprive users of their fair use rights. A legion of supporters, from the ACLU to an alliance of computer programmers, backed her position.

The Second U.S. Circuit Court of Appeals in New York agreed, too, but it still ruled against Corley. In November 2001, the 11-judge panel wrote that DeCSS was illegal under the DMCA because it circumvented technology used to control the use of DVD content. "Even though computer code qualifies as speech, the DMCA regulates only its content-neutral function—the quality that allows the code to instruct a computer to perform," wrote the court. No matter what other information DeCSS might convey, the court said, the government has an interest in restricting its "non-speech" aspect to protect copyright holders, such as the motion picture industry.

In a parallel case, a motion picture industry trade group, the DVD Copy Control Association, sued Johansen, the Norwegian teen who had written and posted DeCSS on the Internet. In that suit, brought under trade secrets statutes rather than DMCA, the court favored the Norwegian teen, ruling that the First Amendment guaranteeing free speech trumps trade secret protections. California's Sixth District Appeals Court deemed the code a constitutionally protected "written expression of the author's ideas and information about decryption of

DVDs." Critics of the DMCA had complained that content encryption such as CSS violated copyright law because it was designed to prevent casual copying and fair use rather than prevent infringement.

The dispute continues. In January 2003, in a similar case, the Supreme Court, with Justice Sandra Day O'Connor writing for the majority, threw out an emergency stay that barred Matt Pavlovich from posting DVD decryption programs on the Internet. Pavlovich's attorney had argued that the stay was unnecessary because such decryption programs already were widely available online.

Felten v. RIAA

Using the DMCA's heavy firepower, content providers continued to play hardball, but their strategy sometimes backfired, especially in the court of public opinion. That was the case in September 2000, when the recording industry launched its Secure Digital Music Initiative (SDMI). The industry invited the public to attempt to break several watermarking security technologies SDMI was testing to protect digital distribution of music. To download the encrypted files to test, contestants agreed to a "click-wrap" contract—a series of web screens and "I Agree" buttons. Under the agreement, SDMI offered to $10,000 for each successful attack. But to collect the money, contestants were bound by the click-wrap contract to assign the intellectual property rights of their labors to SDMI. They also had to promise not to disclose any details of a successful attack. It was only a matter of weeks before Edward Felten, a Princeton University associate professor, and a team of researchers from Princeton, Rice University, and Xerox, found vulnerabilities in SDMI's watermarking. Instead of claiming the reward, Felten planned to present the group's work at the Fourth International Information Hiding Workshop in Pittsburgh.

The Recording Institute Association of America (RIAA), the lobbying arm of the music industry, demanded that Felten withdraw his paper or face prosecution under the DMCA. Felten buckled. But he and the Electronic Frontier Foundation also used the event to test the DMCA's constitutionality. They sued the RIAA, also naming the Justice Department as a defendant, arguing that the SDMI's demands had violated Felten's right to free speech and would have a chilling affect on academic research.

The RIAA backed down, saying it never intended to sue Felten, and moved to have the case dismissed. In its own motion to dismiss, the Justice Department said that Felten's research had never been threatened by the DMCA because the act allows research that is designed not to circumvent, but to study and bolster access control measures. The DMCA, Justice concluded, would no apply to Felten's conduct, and "any fear of an immediate prosecution would be unreasonable."

In November 2001, after 25 minutes of debate, Judge Garrett Brown of the Federal District Court in Trenton, New Jersey, threw out Felten's case, siding with the RIAA's argument that it had never intended to pursue any action. In February 2002, the Electronic Frontier Foundation decided not to appeal.

U.S. v. Elcomsoft

In the summer of 2001, agents from the Federal Bureau of Investigation swooped into a Las Vegas hotel for an intellectual property sting. Their unlikely culprit: a Russian programmer named Dimitry Sklyarov, who had flown to the United States for a programming trade conference, Defcon, to present his latest idea. Sklyarov's program allowed users of Adobe Acrobat e-books to circumvent the program's encryption so they could convert their legally purchased digital books to a more versatile PDF format that could be loaded onto a home computer where the digital text could be read more easily or printed out.

But under the DMCA, paper books and e-books are not created equal. The act trumped longstanding rights to fair use and first sale that buyers of old-fashioned bound books enjoyed. Software supplier Adobe complained to the Department of Justice that Sklyarov and his Moscow-based employer, ElcomSoft Co., had violated Section 1201 of the DMCA by writing code that cracked Adobe's BookReader encryption. The program, Adobe complained, could be used for illicit purposes, such as making illegal copies of books that had not been purchased.

Skylarov's case hit the front pages of major newspapers, including the *New York Times* and *Wall Street Journal*. His arrest and month-long imprisonment captured the public's attention and fueled a broader awareness of the DMCA's far-reaching implications. A young husband and father of two small children, the Russian programmer made for a good news story. Sklyarov was indicted with four counts of circumventing digital rights management technology and a single charge of conspiracy to traffic in a circumvention program. He faced up to 25 years in prison and a fine of up to $2.25 million. ElcomSoft Co. was threatened with a $2.25 million fine. To win his freedom, Sklyarov reluctantly agreed to testify against his employer.

His testimony wasn't enough to sway a jury. On December 17, 2002, a jury acquitted Elcomsoft of software trafficking. But thanks in part to the jury, the ruling didn't put much of a dent in the DMCA. In an interview with Cnet News.com, jury foreman Dennis Strader said jurors agreed ElcomSoft's product was illegal under the DMCA. But they acquitted the company anyway because the statute was "confusing" and jurors believed the company had not intended to break the law. Jurors noted that Elcomsoft had withdrawn its product immediately after being notified that it might be illegal. And during the two-week trial, an Adobe engineer acknowledged that his company did not find any illegal ebooks online and thus couldn't prove that ElcomSoft's program had been used to make illegal copies of ebooks. The DMCA emerged unscathed.

Peer-to-Peer: Industry's Worst Nightmare

Peer-to-peer networks—known also as P2P exchanges or file-swapping networks—are one of the Internet's greatest innovations. Simply put, P2P networks are an easy way to find a data file on the Internet, such as a song, picture, or document. What makes peer-to-peer networks exceptional is that they search and find files that are stored not on a central server, but on another individual's personal computer; hence the description "peer-to-peer." To participate in a P2P network, computer users must install software that acts both as a client that can initiate search requests and receive data, and as a server that can respond to queries and transmit data. As Alan Zeichick noted in the December 2000 issue of *Red Herring*, in a true P2P system, there is "no central index, no meeting point, no single point of failure." Only millions of individual computers using a common program to talk to one another.

Napster

Napster, an early P2P network, rocked the music industry in 1999 when millions of people co-opted it to trade digitized music online. The brainchild of a teenaged college dropout named Shawn Fanning, Napster's software facilitated direct connections between individual users, for example, between a music fan who wanted a particular song and another fan who had that song stored on his hard drive. Downloaded files never actually passed through Napster's computers, which serve only as an index and meeting point for users. Although Napster soon would become synonymous with P2P, it was not completely decentralized. At its peak it relied on nearly 200 of its own servers to facilitate swaps, making it an attractive target for lawsuits and hackers. Shutting down Napster's servers would stop the network and its users cold.

In January 1999, Napster was like any other under-funded Internet startup. Fanning conceived the system in his Northeastern University dorm room, inspired in part by a friend who had complained that he wanted to download songs, but it was hard to find what he was looking for in the vast Internet anarchy. Existing sites that offered digital music downloads in the form of compressed files known as MP3s were run as hobbies by music fans or techies and were unreliable and often outdated.

Fanning had an idea. In testimony to the Senate Judiciary Committee on October 9, 2000, Fanning said he envisioned a communal system that would be "affirmatively powered by the users, who would select what information they wanted to list on the index." When a user logged on to the application, their list would be updated. When they logged off, their files would disappear from the index. Thus the number of song files available at any given time on a site would depend not on the site itself, but on its users.

At first, Fanning simply wanted to prove that the concept would work. But 30 beta versions of the Napster system he passed around to friends in the summer of 1999 spread over the Internet like a virus. By August, before Napster was even incorporated as a legitimate business, the State Department's networks were clogged because so many government employees were downloading MP3 files using Napster. Relying on financial support from family and angel investors, Napster opened shop in September 1999. By December, Hofstra University was so overwhelmed with Napster traffic that administrators sought to block student access to the network. Dozens of universities followed Hofstra's lead, despite student protests.

That month, RIAA sued Napster for copyright infringement, demanding damages of $10,000 for each download. Heavy metal band Metallica sued the following March. In addition to targeting Napster, Metallica named Indiana University, University of Southern California, and Yale University as defendants, claiming that they didn't do enough to block students from accessing Napster's software.

There were no clear lines dividing pro- and anti-Napster forces. Students joined the fray in a big way in early 2000, when Indiana University student Chad Paulson launched Students Against University Censorship (SAUC), collecting 7,000 signatures from students opposing Indian's efforts to block Napster access on campus. Indiana backed down, only to be named in Metallica's litigation. Paulson later flip-flopped, disavowing SAUC. Meanwhile, 300,000 Metallica fans petitioned Napster to stop facilitating the theft of the band's music. Even the music industry was divided. In April 2000, Napster sponsored rock band Limp Bizkit's tour. Limp Bizkit and many other musicians, particularly smaller, less-commercial artists, saw peer-to-peer technology and the Internet as a way around the strict commercial and artistic controls of the profit-minded record labels.

Napster's scalability was the music industry's worst nightmare. By using the Internet to reach millions of users, it had transformed a commonplace but heretofore largely ignored practice—the private sharing and occasional copying of music between friends and fans—into a staggering global exchange that threatened to take a big bite out of the recording industry's bottom line by making nearly any recording available to anyone, anywhere, for free. Napster also was a headache for network administrators, particularly on college campuses, where students were P2P's biggest users. File-swapping ate up huge amounts of bandwidth and slowed high-speed connections. By 2000, Napster had some 1.5 million simultaneous users who had logged 70 million downloads. That year, the recording industry had tallied its losses at some $300 million, a number that would more than double in the next two years as Napster clones surfaced.

Napster's ranks continued to grow. Within a year, despite the litigation and a mixed reception from the public and the press, it was one of the most-trafficked sites on the Internet. By the time the Ninth Circuit Court of Appeals heard oral arguments in the case in October 2000, Napster boasted 50 million users. As the case was going to trial, Napster settled with media giant Bertelsmann BMG, one of the plaintiffs. Bertelsmann, which owned rights to popular artists such as Carlos Santana, agreed to drop its case while it worked out subscription or pay-per-download service with Napster.

Napster argued before the Ninth Circuit Court of Appeals that its operations were protected under the 1992 Audio Home Recording Act, which allowed consumers to make home copies of music for their personal use. But the court had little sympathy for Napster, declaring it liable for contributory and vicarious copyright infringement. The decision chipped away at the Supreme Court's opinion in the 1984 Betamax case, where a 5-4 majority held that movie studios could not outlaw a technology—videocassette recorders in that case—just because the technology could be used to illegally copy content. The court ordered Napster shut down in February 2001. Napster went dark and declared bankruptcy, but its legacy lives on. Bertelsmann bought the Napster name and in November 2003 relaunched the site as a legal, pay-to-play network.

Napster Redux

The industry's victory didn't end the war over peer-to-peer technology. Before the Napster battle was over, more than a half dozen new and improved Napster imitators had taken up the flag, including KaZaA, Grokster, and Morpheus. RIAA estimated in 2003 that some 3 billion songs were illegally downloaded from these sites every month. Music isn't P2P's only commodity. Games, movies, and software are making the rounds, too. Federal prosecutors are fighting the battle on the criminal front, sending a message to hackers and pirates by making examples of their peers. In May 2002, the Justice Department charged Robin Rothberg, a 34-year-old Boston area computer consultant, with conspired to infringe the copyrights on thousands of software programs worth more than $1 million. Rothberg pled guilty and was sentenced to 18 months in prison for his role as leader of Pirates With Attitude, an online software piracy group. The Pirates, like hundreds of similar cyber-based groups, did for computer software what Napster did for music, swapping games and financial programs on line. No money changes hands, and the groups are mostly hobbies for members. Despite the lack of financial incentive, under the DMCA, such "recreational" piracy is a criminal offense punishable by as much as three years in jail. Rothberg and 16 other Pirates With Attitude were the first group to see jail time under DMCA.

DMCA critics complain that the law is being used to thwart legitimate and legal uses of technology and give broader control to copyright owners such as movie studios and record labels. They say digital theft is fought more effectively by going after big offenders such as overseas piracy rings which, unlike small timers like Rothenberg, take an appreciable bite out of corporate profits. Content providers would be better served, too, if they charged fairer prices for their products, say consumer groups. Audiophiles point to the high price of compact discs—which cost about $3 to make but sell for about $18—as a driving factor behind piracy, especially among younger consumers. And they claim that the music album as an artistic format is no longer recognized by the industry, nor is it commercially viable, thanks to the industry's domination by studio-bred artists with short-lived recording careers that rely on a handful of top commercial hits instead of an expansive songbook of work. Compact discs brought the demise of the single record, but the Internet and music fans are clamoring to bring it back.

Film and music companies counter that they have a responsibility to artists and shareholders, as well as every right to aggressively protect their intellectual property. Music labels blamed digital piracy for a double-digit drop in compact disc sales in 2002. Yet even as they play host to widespread file-swapping, P2P networks are proving extraordinarily difficult to topple. KaZaA, among the biggest, is a case in point. The program's corporate owner, Sharman Networks, is incorporated on the island nation of Vanuatu. The company is run out of Australia. Its servers are located in Denmark. The source code for its program was last seen in Estonia, and its developers are thought to be in hiding in the Netherlands. Geography is only one hurdle. KaZaA's file-swapping program, although used predominantly by online music swappers, has any number of legitimate uses. Echoing the Betamax claim, the company claims that software that makes it easy to for users to violate copyrights isn't itself illegal.

Recording studios doggedly continue litigation. But frustrated in their attempts to shutter P2P networks, music and movie studios are tackling digital piracy on new fronts. They are suing universities that enable student file-swapping by dedicating servers and high-speed pipes to the activity. In early 2003, the industry targeted corporate America, putting the CEOs of every Fortune 500 company on notice: businesses would be sued if their employees were using company networks to swap files.

A few months later, the RIAA cast an even wider net. Between April and August, the industry used instant messaging to threaten an estimated 4 million file swappers that they could face fines and jail time for illegally downloading music. Industry was taken to task for "spamming" even innocent P2P users, and KaZaA and other sites moved quickly to block the RIAA's messages. The RIAA's next move: taking advantage of the DMCA's relatively low standards of just cause to issue some 1,100 "informational" subpoenas to suspected file-swappers in preparation for a flood of lawsuits against individual downloaders. In September 2003, the industry sued 261 file-swappers, many of them minors. Music fans, legal analysts, and the press have castigated the industry's take-no-prisoners approach—it is, after all, suing its own customers. But the strategy seems to be working. The RIAA began settling cases the day after they were filed, with some defendants paying fines in the five figures. Traffic on P2P sites slowed markedly while legal music download sites such as Apple's iTunes, selling individual songs for 99 cents, watched business boom.

Legitimizing Peer-to-Peer

Indeed, hampered as they were by security concerns, recording studios only belatedly recognized the Internet's potential as a distribution system. It wasn't until mid-2003 that major studios began teaming up to offer subscription-based services that allow music lovers to download their favorite songs in return for a monthly subscription fee or per-song charge. But fears over privacy continue to make the system imperfect. Most services are sticking with "tethered" downloads that license a user to listen to a song a limited number of times, and only while his or her computer is connected to the Internet. Four years after the Napster saga, systems allowing outright purchase of music from the Internet, complete with the ability to copy songs onto compact discs, remained in the formative stage. The most successful, Apple's popular iTunes, boasted 13 million song sales, at 99 cents a pop, in its first six months. By the end of the year, it had licensed 400,000 songs and 5,000 audiobooks. But thanks to byzantine rules governing music and book licensing, even that popular service prevents legally purchased music from being played from a computer hard drive if it's taken outside of the United States.

Hollywood, already scarred by China-based factories churning out thousands of illegal movie copies on DVD, fears that napsterization of the film industry is next. Industry executives say widespread deployment of high-speed Internet connections, as well as the roll-out of high-definition television, will open the door to new waves of piracy. Studios are encrypting music and movie discs so they can't be uploaded, converted to MP3 format, or even played on a computer's CD-ROM. The Motion Picture Association was embarrassed into installing identifying tags on its DVDs when *Lord of the Rings: The Two Towers*, Chicago, and other Academy Award–nominated movies wound up for sale on the streets of China. Hollywood studios, as part of a public relations campaign to win votes for their award nominees, had distributed to critics hundreds of DVDs that found their way into illegal markets.

A Library Takes the Lead

Billions of dollars are at stake in the copyright wars, but a little-known and hitherto non-influential federal agency is at the forefront of managing digital rights. In passing the Digital Millennium Copyright Act, Congress's goal in part was to encourage copyright owners to use the Internet and other digital distribution methods to make their creative works available to the public. Lawmakers knew the DMCA would require routine revisions if it was to keep pace with rapid technological advances. They also recognized that the law wasn't perfect: Entities with a vested interest in protecting control over their works or their profit margins might abuse the DMCA's anti-circumvention provision, which makes it a crime to bypass any "effective technological protection measure," and thus "unjustifiably diminish" public access to copyrighted material. Indeed, Section 1201 as written makes no distinction between circumvention for theft, circumvention for personal use (for example, a music lover who wants to convert a legally purchased music CD for use on an MP3 player), or circumvention for academic research or security (a software designer who wants to pick apart a program to find ways to improve it).

In a 2000 report, the National Academy of Sciences admonished Congress for this "imprecise provision" of the DMCA. "Is a measure that can be circumvented by anyone who has successfully completed a freshman-level college course in computer science an 'effective technological protection measure'? What is the threshold for 'effective'?" The Academy also noted that the DMCA doesn't recognize many legitimate reasons for circumvention, such as reverse engineering. Under the law, legions of academic researchers and computer industry professionals could be regarded as criminals: "A copyright owner might need to circumvent an access control system to investigate whether someone else is hiding infringement by encrypting a copy of that owner's works, or a firm might need to circumvent an access control system to determine whether a computer virus was about to infect its computer system."

Lawmakers tried to address such concerns by instructing the Library of Congress's Copyright Office to review the DMCA every three years. The provision granted the Copyright Office the authority to make exceptions to the DMCA's anti-circumvention rules and, in theory, protect consumers from unreasonably aggressive digital rights management schemes that might deprive the public access to the creative works. In October 2000, after the first triennial review of DMCA regulations, the Librarian of Congress, at the recommendation of the Registrar of Copyrights, exempted two classes of works from Section 1201 of the DMCA: Compilations of lists of websites blocked by software filters, and literary works (including computer programs and databases) protected by controls that, because of malfunction or obsolescence, failed to permit access (see Box 3 for additional provisions of the DMCA). At the time, the Copyright Office weighed in against fair use advocates, noting that "there is no unqualified right to access works on any particular machine or device of the user's choosing."

Content providers utilize many methods to protect and manage their property. A sampling of these methods is provided in Box 2.

Box 2: Digital Rights Management

Licensing—Discards the concept of consumer ownership by selling only the rights to use a product, not the product itself
Watermarking—A form of encryption that indicates that a product is authentic and not an illegitimate copy
Registerware—Requires users to input personal data to make a system operable. Allows content owners to track use
Tethering—Limits how or where a product can be used, or how many times it can be copied
Piracy Tax—A tax on consumer electronic equipment or software meant to make up for losses from piracy

Box 3: Other DMCA Provisions

Limits caching by Internet service providers and search engines such as Google
Prohibits websites from linking to DMCA-illegal material
Requires Internet service providers to act "expeditiously" to block or otherwise respond to illegal content or activity
Grants leeway to libraries, which are allowed to make three digital copies of a particular holding but may not circulate the copies outside library premises (with the word "premises" as-yet undefined)
Requires compulsory licensing by recording industry to music webcasters, with fees arbitrated by the Library of Congress
Allows bypassing of technological copyright protection under very limited circumstances

DMCA: Taking a Second Look

While the music industry was making the most of the Digital Millennium Copyright Act to fight piracy, it was just beginning to dawn on Hollywood that it might be P2P's next big victim. Widespread rollout of broadband made downloading DVDs much easier. And P2P was just the beginning. New products, such as TiVo, SonicBlue, and ReplayTV were changing the way consumers tuned into prime-time television. And the slow but inevitable adoption of digital broadcast technology nationwide were raising the specter that movies and television programs would be napsterized the way the music industry had been.

But while network executives and filmmakers were having nightmares, consumers, Congress, and the electronics industry were having second thoughts. The DMCA was having a chilling effect on new technology, the electronics industry claimed. Instead of hiding behind Congress and the courts and dictating how content could be used, the entertainment industry needed to face reality—consumers were demanding new ways to receive and enjoy music and programming. Technology could deliver on that demand but for the entertainment industry, which had been slow to try new business models and adopt novel methods of distribution. The lines were sharply drawn in 2002, when the entertainment industry went to Congress with more demands for copyright protection, including a list of technology industry mandates. Despite the industry's deep pockets and broad political influence, it would be an uphill battle. Hollywood met with fierce resistance from consumers, cyber-libertarians, academics, and most important, the massive technology industry. It became what Stanford's Larry Lessig called a war over rival visions of the digital future.

By mid-2002, Silicon Valley and Hollywood were locked in a titanic struggle that played itself out on Capitol Hill. On one side were entertainment giants such as Walt Disney Chief Executive Officer Michael Eisner, News Corp. Chief Operating Officer Peter Chernin, Motion Picture Association of America President and master lobbyist Jack Valenti, and AOL Time Warner CEO Richard Parsons. They accused the consumer electronics industry of condoning digital thievery and stalled efforts to protect television from illegal copying. Until sufficient piracy protections were in place, no one in Hollywood would risk making high-quality digital programming. High-definition television would be on hold, and consumers stuck with old-fashioned analog sets would be denied the wonders of digital entertainment.

Hollywood's point man in the Democratic-led Congress was Senate Commerce Committee Chairman Ernest F. Hollings (D–SC). The industry asked Hollings to introduce legislation that would mandate an end-to-end digital rights management system, giving computer and consumer electronics makers a deadline for developing anti-copying "policeware" and installing it on all newly manufactured products, including televisions, computers, camcorders, CD burners, and MP3 players. That was just part of the entertainment industry's short but demanding wish list. They wanted Congress to protect the theft of over-the-airwaves digital broadcasts by mandating that all television sets include a way to detect invisible digital "flags" embedded in programs, which in theory would prevent a show from being uploaded and distributed over the Internet. Hollywood also sought a standardized watermarking system to protect analog broadcasts and technological fix to the still-pressing problem of file-swapping. Hollings and the industry's top House allies, Representatives W.J. "Billy" Tauzin (R–LA), and John

Dingell (D–MI), the chairman and ranking member of the Energy and Commerce Committee, directed the Federal Communications Commission to look into ways to thwart online piracy. The entertainment industry was well-connected and had a compelling story to tell. At every turn, entertainment lobbyists barraged lawmakers with visual examples of outright theft. Valenti screened excerpts of a pristine stolen recording of the blockbuster film *Gladiator* for a Senate audience.

But this time, the Hollywood lobby had met its match. Intel CEO Craig Barrett, Microsoft founder Bill Gates, Apple's Steve Jobs, and a pack of tech goliaths that included Cisco Systems, Philips Electronics, Zenith, and Hewlett-Packard, mobilized to fight a one-size-fits-all mandate on piracy protections, fretting that it couldn't possibly envision and accommodate future technologies and would chill innovation. This already powerful group—an industry 20 times the size of Hollywood—had company. While the tech industry pulled strings on Capitol Hill, Consumers Union, the American Library Association, and a loose-knit gang of cyber-libertarians led by Larry Lessig and the Electronic Frontier Foundation helped rally grass-roots opposition to Hollywood's idea, spurring more some 3,300 people to write letters opposing the Hollings plan. Every technology association in Washington, including the Information Technology Industry Council, TechNet, and the Association of Computing Machinery, came out against content providers. Consumer groups warned that the Hollings proposal would jeopardize fair use by limiting a consumer's ability, for example, to email home video clips to friends. Finally, the anti-Hollywood crowd turned to Congress's "cyber senator", Senate Judiciary Committee Chairman Patrick Leahy (D–VT), a technology champion whose panel had jurisdiction over copyright issues. Leahy and his allies issued dire warnings about Congress getting involved in picking technology winners and losers. The Hollings bill died a slow death.

Hollywood and Silicon Valley continue to work on copy protections under the auspices of the Copy Protection Technical Working Group, a collection of more than 100 engineers who meet regularly to collaborate on piracy protections that will work without harming the technology sector or consumers. The group had one early victory, a way to scramble and encrypt programming to prevent it from being copied, but progress since then has been slow as infighting continues.

IP and Antitrust

While policy makers continued to debate the strengths and shortcomings of the Digital Millennium Copyright Act, the law's ban on circumvention of content-protection technology was beginning to have unintended consequences. The most notable was its strict prohibitions on circumvention, which had given manufacturers a tool to stifle competition in ways never before imagined. That was the case with Kentucky-based Lexmark International had devised a novel way to force buyers of its computer printers to use only Lexmark replacement toner cartridges. Lexmark, a leading manufacturer of printers and components, equipped its cartridges with computer chips that send an authentication sequence—a "secret handshake"—to a chip in the printer. If a user installs a generic toner cartridge, sans chip, the printer won't work. The microchip self-destructs when the toner is empty, thus preventing Lexmark cartridges from being remanufactured or recycled by its competitors.

Stanford, North Carolina-based Static Control Components found a way to mimic Lexmark's chip and began selling after-market toner cartridges for Lexmark printers. Lexmark sued in U.S. District Court in Lexington, Kentucky, in December 2002. Among the company's claims: the programming in its microchips were protected under the DMCA, and Static Control illegally had circumvented the chips' protection measures.

The case roiled the $40 million computer printer industry, where printer cartridge sales make up the bulk of revenues. Lexmark said it is trying to protect its investment in research and development, and claimed that its printers use copyrighted software to control various printer functions, including paper feed, paper movement, and motor control. Access to the copyrighted programming is protected by a technological measure—the authentication sequence from the toner cartridge. Static Control's computer chip circumvents that technological measure and as such provides unauthorized access to the printer's copyrighted software. And circumventing a technological measure to gain access to a copyrighted work violates the DMCA. Static Control won support from industry trade groups such as the Computer & Communications Industry Association, which feared that a ruling in Lexmark's favor could make reverse engineering all but illegal or lead to widespread copyrighting of the most basic programming language like that used in the Lexmark chip. Other groups, such as automotive parts manufacturers, expressed concern that a win for Lexmark could wipe out the aftermarket parts industry altogether if manufacturers began adopting chip-enabled technologies on a widespread basis. In February 2003, Lexmark won a restraining order against Static Control, which was blocked from selling its Lexmark-compatible cartridges until the case was heard.

With its case awaiting trial, Static Control won an unexpected victory in October 2003. The Copyright Office, as part of its second tri-annual evaluation of the DMCA, refused to grant Static Control an exemption to Section 1201 so it could reverse-engineer Lexmark's chip so it could make compatible printer toner cartridges. In her recommendation to the Librarian of Congress, Copyright Registrar Marybeth Peters found that Static Control didn't need a 1201 exemption because reverse-engineering a product for the purposes of making another product work with it—the concept of interoperability—already is protected under the DMCA. "Interoperability necessarily includes...concerns for functionality and use, and not only of individual use, but for enabling competitive choices in the marketplace," Peters wrote. Peters premised her ruling on the fact that Static

Control had easy and legal access to Lexmark's chip—because Static Control had purchased a Lexmark printer. Under the DMCA, access is a necessary requirement of legal reverse-engineering for the purpose of interoperability. Because Lexmark's programs "were available in the regular toner cartridge, these programs were claimed to be 'readily available' to Static Control," Peters wrote.

Just two weeks later, DMCA critics won another round. In Chicago, the Chamberlain Group, a manufacturer of mechanical garage-door openers, had sued Skylink Technologies, which made remote controls that worked with Chamberlain's garage-door openers. Chamberlain's machines contain a security feature that ensures a door will open only when the machine receives certain software codes from its remote. This "rolling code" technology was designed to protect homeowners from "code-grabbers" who might steal a remote's transmission and use it to gain illegal access to the garage. Skylink remotes circumvent that technology, Chamberlain claimed, thus violating the DMCA. In November 2003, Judge Rebecca R. Pallmeyer of the Northern District of Illinois District Court ruled that Skylink's universal garage door clicker didn't violate the DMCA because homeowners who bought Chamberlain garage doors had legal access to its remote programming. "Under Chamberlain's theory, any customer who loses his or her Chamberlain transmitter, but manages to operate the opener either with a non-Chamberlain transmitter or by some other means of circumventing the rolling code, has violated the DMCA," Pallmeyer wrote. "Transmitters are similar to television remote controls in that consumers of both products may need to replace them at some point due to damage or loss, and may program them to work with other devices manufactured by different companies. In both cases, consumers have a reasonable expectation that they can replace the original product with a competing, universal product without violating federal law."

Manufacturers had tried to use copyright law to quash competition or stifle third-party add-ons before the DMCA came on the scene. In a landmark 1992 case, Sega Enterprises, maker of a popular game console, sued a rival game maker for infringement, arguing that the game maker embedded Sega's copyright-protected software into its games so they would work on Sega's machines. The Ninth Circuit Court of Appeals held that copying software code to create interoperability is fair use, and the game-maker won. The courts so far have yet to explicitly cite or recognize fair use as a defense under the DMCA.

In many cases, the courts haven't even gotten a chance to weigh cases, thanks to a DMCA provision that gives content owners broad powers to subpoena information and demand that ISPs block offending websites at a copyright holder's demand. In 2002, several retailers, including Wal-Mart, Target Stores Inc., Best Buy Co., Staples Inc., OfficeMax Inc., Jo-Ann Stores Inc., and Kmart Corp., took aim at FatWallet.com, a bargain hunters' website that posted information on current and upcoming sales. The retailers claimed that FatWallet had stolen their intellectual property—the prices the stores assign to items. FatWallet disputed that prices could be copyrighted, but the site went dark because its founder couldn't afford a long court case. Dow Chemical Co. used the DMCA to shut down a Web site that parodied the company. And Apple Computer cited the DMCA to stop one of its dealers from producing and selling software that allowed Apple's new DVD-burning technology to be used on earlier models of its Macintosh computers.

Under DMCA, intellectual property has become a tool for shaping markets rather than simply generating revenue.

Shrink Wrap

E-publishing promised to be the biggest thing since the printing press. It would cut costs and boost margins for imprints by eliminating the high cost of paper. It was a way to boost magazine circulation by allowing quick and cheap delivery in a global market. It held out the promise of new revenue stream by allowing advertisers to insert multi-media advertising in a publication's digital "pages." Expensive encyclopedia sets were replaced by full-color, interactive digital volumes. Hospital personnel carried their digital pharmaceutical guides on Palm Pilots as they visited patients. Lawyers toted entire law libraries on their laptops. Writer Stephen King jumped on the bandwagon in 2000 with *Riding the Bullet*, the first novel to be published exclusively in digital format. In 2001, book publisher and retailer Barnes & Noble teamed up with e-book publisher Adobe to launch an e-book store on the Internet. Concepts associated with technology migrated to the print world, where publishers of old-fashioned paper books began experimenting with shrink-wrap licenses. Now many trade groups and professional associations shrink-wrap their printed membership directories and other print publications to prevent them from being digitized and circulated online.

But e-books in their many forms blurred the line between books and software, upending the long-held notion of first sale. Was the CD version of Encyclopedia Britannica software or a book? For publishers, the answer was simple. Whereas paper books presented huge barriers to widespread theft, including the cost of copying and distribution, e-books could be duplicated and sent anywhere for almost nothing. In 1998, the publishing industry, led by the Association of American Publishers, pushed legislation that would have created a first sale doctrine for the transmission of digital works. The proposal would have required owners of digital books to delete their copy if the work were transmitted to someone else.

The complicated plan never gained traction in Congress. With their legislative options limited, publishers turned to technological solutions. Developers of e-book readers such as Microsoft and Adobe incorporated encryption technology that, depending upon the publisher's wishes, forced e-book readers to use a certain operating system to read their e-books, or limited their reading to a certain number of computers. ElcomSoft's programming picked those digital locks, enabling e-book purchasers to read their books on Linux operating systems, print out pages, and use text-to-speech processing.[1] Microsoft's program limited e-book readers to two separate machines, such as a home desktop and a laptop. Adobe's Glassbook (later renamed Acrobat eBook Reader Plus V2.0), the first platform for King's novel, wouldn't allow the writer's fans to print out the book's pages. Such approaches only alienated readers. Purchase of a new home computer, for example, could wipe out a reader's access to the novel he had purchased. And readers who preferred reading from a print out instead of curling up in bed with a PC simply were out of luck. Not surprisingly, e-books, with the exception of references and other commodity works, have been slow to catch on. Barnes & Noble shuttered its e-book operation less than three years after its launch.

Open Source Code

"Alchemists turned into chemists when they stopped keeping secrets."

—Eric Raymond, author, *The Cathedral and the Bazaar*

At the root of every computer program is its source code—the language that tells a computer how to do something. For most applications, the code is a closely guarded secret, one that corporations must protect to ensure the fidelity of their products and to prevent their work from being replicated freely.

Makers of such proprietary software, including Microsoft and Apple, sometimes share sections of source code with their larger clients on a need-to-know basis. But that access remains the exception to the rule. Complex programs take great time and expertise to create, which is why software manufacturers erected multiple layers of protection around their products. They've found that copyright protection no longer is enough. Like recording studios, software companies are installing digital protections into the programs they sell to prevent consumers from making illegal copies or using programs on multiple computers. And they are pushing legislation to make shrink-wrap licenses legally binding.

But by the end of the 1990s, the very notion of proprietary software was under heavy fire from an eclectic, loosely knit, non-profit-minded clutch of IT professionals and students who share a utopian vision, where the exchange of ideas and knowledge is completely unfettered by notions of ownership. In the information technology sector, this open-source movement challenged traditional concepts of capitalism and intellectual property.

---

[1].   See *http://www.eff.org/IP/DMCA/US_v_Elcomsoft/us_v_elcomsoft_faq.html#ChargedWith* and *http://www.adobe.com/aboutadobe/pressroom/pressreleases/200108/elcomsoftqa.html*.

A weak economy, combined with deep public distrust of the omnipresent Microsoft goliath, has propelled a boom in corporate and government use of open source software, particularly the Linux operating system. In 2003, what had been a subterranean movement of techno-purists was poised to erupt into a full-scale economic revolution. By giving away its source code, Linux threatened to bring down business models that relied on intellectual property protections to generate revenue and profit. Corporations under pressure to cut operating costs cast about for ways to reduce their computing bills and discovered a low-cost solution in Linux.

Morgan Stanley was among them. The company had its eye on the bottom line in 2003 when it replaced 4,000 high-powered servers with much cheaper Linux-based systems. The firm expects the switch to generate $100 million in savings by 2008. Several tech industry goliaths, including Intel, Oracle, and Dell, also threw their weight behind open source. IBM was among them, devoting $1 billion and 250 engineers to retooling its software and computers to work with Linux. "We can't tell our customers to wait for Linux to grow up," IBM Vice President Robert LeBlanc said. By the first half of 2002, 15 percent of the mainframes shipped by Big Blue ran Linux, and the company had sold $160 million worth of Linux servers. By 2003, IBM had 4,600 Linux customers. A January 2003 survey by Goldman, Sachs & Co. found that 39 percent of large corporations used Linux. Analysts expect Linux's share of the $51 billion market for computer servers to reach 25 percent by 2006, up from almost nothing in 1999.

Governments will be the next big consumers of open source programming. Germany is funding an initiative to develop open-source security software. The European Union is considering underwriting development of a common pool of open-source software for use by its member governments. Several large cities in Brazil give procurement preference to open source. Peru is considering legislation to require all public agencies to use open-source software. Singapore gives tax breaks to companies that use Linux rather than proprietary operating systems. Taiwan is trying to wean the entire country from Windows by subsidizing development of open source software, and training hundreds of thousands of people to use Linux. And China is channeling big government contracts to Red Flag, a private developer of Linux-related software.

Open-source preferences have been slower to gain traction in the United States, but efforts are under way to give it government's blessing. In 2002, California state lawmakers mulled the Digital Security Software Act, a bill requiring state agencies to use open-source software. And under a regulation that requires the fruits of federal research to be made easily accessible to industry, some software designed under federal programs is being released under a General Public License (GPL), which prohibits anyone from making a profit off the program. NASA, Sandia National Laboratories, and the Defense Advanced Research Projects Agency have developed or funded development of open-source code.

But open-source software that costs nothing to obtain and use isn't the same as software that's in the public domain. Many open-source systems, including the widely used Linux, are copyrighted and licensed under terms of a General Public License. Although anyone is free to download or copy GPL software, the license restricts how it may be used. For example, programs that grow out of modifications made to GPL software must be licensed under the GPL, too. Thus, any software containing even modest amounts of GPL code itself falls under GPL restrictions—the code becomes viral.

GPL was the brainchild of the Richard Stallman, a programmer formerly with MIT's Artificial Intelligence Lab. To promote the open source movement and ensure that open-source systems would remain free to the public, Stallman founded GNU, a non-profit organization, to hold and administer the General Public License.

"Free" has multiple meanings, with cost being secondary. "When we speak of free software, we are referring to freedom, not price," reads the GNU license. "Our General Public Licenses are designed to make sure that you have the freedom to distribute copies of free software (and charge for this service if you wish), that you receive source code or can get it if you want it, that you can change the software or use pieces of it in new free programs; and that you know you can do these things." GNU's main driver is to keep free software from falling under proprietary patents: "Any patent must be license for everyone's free use or not licensed at all." The license disclaims any warranty from defect or malfunction, a provision open-source critics use to bolster their case that free software is sub-par to propriety products.

Indeed, GPL's critics predict that the open-source movement ultimately will chill innovation by removing financial incentives to build on open-source code or cross-fertilize for-profit programs with ideas from GPL software. "This condition is not as innocuous as it might first appear," says David S. Evans, senior vice president of National Economic Research Associates. GPL "cuts off an important avenue of innovation in products of enormous importance to all industrialized economies," Evans says. "Since most modern programs have roots in earlier generations of software, the mere presence of lots of GPL software in the computing environment generates uncertainly about property rights."

Case in point: Programmers at NVIDIA, a developer of computer graphics devices, inadvertently used a few lines of GPL code in some new software. When the copyright holder demanded that NVIDIA honor the GPL and release the source code for the whole program, the company had to go back to the drawing board and reprogram its software.

Software's dependence on network economies, in which the value to each user grows with the number of users, also makes the sector particularly vulnerable to changes in how intellectual property is defined. Adoption of a patchwork of open source systems could balkanize software markets in which efficiency demands winner-take-most outcomes. One scenario envisions the return of compatibility problems like those the industry experienced in the mid-1980s, when computer files couldn't communicate with each other or with different machines, and users had to be trained to use software on multiple platforms.

Intellectual property issues pose the biggest risk to the open-source movement. SCO Group, which holds the original patents for UNIX, upon which Linux is based, has created a licensing division and hired super lawyer David Boies to press claims against sellers of Linux. On March 7, 2003, it filed its first case, charging IBM with theft of trade secrets, unfair competition, and breach of contract. The company is expected to next take aim at Linux users.

SCO's lawyers say that as long as Linux development was uncoordinated and random, it posed little or no threat to SCO or other UNIX vendors, because of its poor quality and its status a "fringe" product. IBM's embrace of Linux changed all that. "Prior to IBM's involvement, Linux was the software equivalent of a bicycle. UNIX was the software equivalent of a luxury car," SCO's lawsuit claims. IBM illegally appropriated UNIX code and architecture and dumped a "significant financial investment" into Linux, transforming the system into a luxury car, SCO claims. Big Blue's "scheme" was to tap into growing public acceptance of Linux and use it to compete unfairly: "With ten times more services-related personnel than its largest competitor, IBM sought to move the corporate enterprise computing market to a services model based on free software on Intel processors," SCO's lawyers wrote. IBM extracted "confidential and proprietary information" it acquired from UNIX and dumped it into the open source community, SCO claims. In the process, IBM "deliberately and improperly" destroyed the market value of UNIX.

Open-source proponents find it significant that SCO has not asserted a direct intellectual property claim over Linux on the basis of its ownership of the original UNIX code, which was written by programmers at Bell Labs and ultimately purchased by SCO. Ownership of the original Bell Labs Unix could give SCO proprietary rights to UNIX-derived programs, including Linux.

Open-source movement leaders want corporations to license their patent portfolios for free use with open-source software, pitching the move as public service that would enhance the donor's public-relations efforts. One proposal would allow donor companies to take a tax deduction for their foregone royalties.

What's Next?

In an information-driven society, is unbreachable digital content even possible? A decade of trying hasn't yielded a lasting technological solution, let alone a solution that works and is culturally acceptable. Concerns over content protection and privacy have slowed the roll-out of broadband Internet access, digital television programming and digital broadcast in theaters. Today's digital rights management schemes intrude on free speech and long-accepted cultural and legal notions of fair use and are stoking a consumer backlash. Ownership is giving way to a pay-per-use society. The free exchange of ideas is needed to stoke progress in an information economy, where innovation must be maximized. But intellectual property owners must continue to have financial incentive to create. An April 2002 report from the International Intellectual Property Alliance, which represents groups such as RIAA, MPAA, the Business Software Alliance, and the Association of American Publishers, concluded that copyright industries are a driving force in the U.S. economy, accounting for 5.24 percent of gross domestic product, or $535 billion—an increase of $75 billion from three years earlier. The industry employs 4.7 million workers, or 3.5 percent of total U.S. employment, and accounts for exports of $89 billion, surpassing all other sectors including aircraft, chemicals, and motor vehicles.

The digital revolution requires a broad re-thinking of the notion of ownership. Larry Lessig's Creative Commons, an Internet-based forum that encourages creative thinkers to register their ideas in the public domain, functions on the Jeffersonian ideal that intellectual property can be kept and consumed simultaneously. Established in December 2002, it has attracted a host of creative works that are licensed for free use. But policy makers continue to move in the opposite direction, moving to restrict public use in favor of creating more value for the copyright holder, which more and more is turning out to be a multi-national conglomerate.

The current furor over ownership of ideas in a digital age is based on the old-fashioned industrial assumption that there is a relationship between scarcity and value. Policymakers need to move beyond such thinking while balancing intellectual property rights against the economic imperative to let technological innovation run its course, and the democratic imperative that a free exchange of ideas is needed for the republic to function.

Further Reading

Lessig, Larry, *The Future of Ideas: The Fate of the Commons in a Connected World*, New York: Random House, 2001.

Theirer, Adam, and Clyde Wayne Crews, Jr., *Copy Fights: The Future of Intellectual Property in the Information Age.* Washington, D.C.: Cato Institute, 2002.

Litman, Jessica, *Digital Copyright: Protecting Intellectual Property on the Internet,* Amherst, NY: Prometheus Books, 2001.

Vaidhyanathan, Siva, *Copyrights and Copywrongs: The Rise of Intellectual Property and How It Threatens Creativity on the Internet,* New York: New York University Press, 2001.

Chapter 8

Antitrust

Eric Fisher

$\mathbf{A}$ntitrust law has been an integral part of the development of information technology, particularly over the latter half of the twenty-first century. Microsoft Corporation has been the most visible example of a company that has had to address antitrust issues through a variety of complaints that it has acted in an anticompetitive manner in recent years. In order to best understand how antitrust law has impacted the development of technology broadly, it is important to analyze the Microsoft case with a detailed focus on the underlying allegations of anticompetitive practices. In addition, the history of Microsoft's involvement with allegations of monopolistic practices and the resulting litigation and congressional action must be analyzed with respect to antitrust laws, such as the Sherman and Clayton acts. To develop the historical context of the impact of antitrust law during recent technological development, other companies such as IBM and AT&T will be addressed. Finally, the Microsoft case will be re-addressed with respect to future developments regarding antitrust considerations and the information technology industry.

Defining Antitrust

The hallmark of monopoly power is the absence or ineffectiveness of competitive constraints on price, output, product decisions, and quality. In general, the issue of monopoly power is addressed by defining "the relevant market," and assessing shares in that market. This is at least a beginning guide to the presence or absence of market power, and a way of organizing the facts that one will have to take into account.

Because its purpose is the identification of monopoly power, if it exists, the definition of relevant market should include all those products that reasonably serve to constrain the behavior of the alleged monopolist. Such constraints arise from three sources: substitution by consumers to other products; substitution by producers to other products; and entry of new productive capacity.

Barriers to entry are factors that would prevent entry in the face of supernormal profits. Where there are significant barriers to entry, monopoly power can be present; otherwise it cannot.

The barriers to entry in the present case stem from a combination of economies of scale and network effects, and from the fact that programs written to run on a given operating system will generally not run on others unless considerable expenditures are undertaken.

Like all software, applications programming exhibits substantial economies of scale, because most of the costs come in the creation of the software and are independent of the number of copies that are produced. Hence software developers wish to write for operating systems that have a large number of users.

Network effects arise when the attractiveness of a product to customers increases with the use of that product by others. Indeed, the fact that many applications are written for a given operating system and cannot easily run on other operating systems makes that operating system more attractive to users. Interestingly, the importance of the availability of applications for operating systems networks has prior to this case been unappreciated.

Taken together, these network effects and scale economies create a positive feedback: the more users an operating system has, the more applications will be written for it; the more applications written for an operating system, the more users it will acquire. After this feedback effect has operated for a while, it becomes difficult or impossible for a new operating system to make much of an inroad.

In these circumstances, it is natural for one firm to become dominant in operating systems, acquiring monopoly power. However, the fact that the successful firm has acquired monopoly power with a "natural" barrier to entry does not justify its taking anticompetitive acts to extend that power to another market or, in particular, its engaging in anticompetitive acts that serve to buttress and protect its power in the original market.

Antitrust Law, the U.S. Code, and the Sherman and Clayton Acts

Unites States Code Title 15 provides that Antitrust laws are developed in part to protect trade and commerce against unlawful restraints and monopolies.[1] In addition, the official penalty outlined by the code includes every person who shall monopolize, or attempt to monopolize, or combine or conspire with any other person or persons, to monopolize any part of the trade or commerce among the several States, or with foreign nations, shall be deemed guilty of a felony, and, on conviction thereof, shall be punished by fine not exceeding $10,000,000 if a corporation, or, if any other person, $350,000, or by imprisonment not exceeding three years, or by both said punishments, in the discretion of the court.[1] The most critical element regarding antitrust law that can be identified in relation to the United States Code and Microsoft relates to what steps are necessary to find a judgment of an antitrust violation.

The code provides that:

> A final judgment or decree rendered in any civil or criminal proceeding brought by or on behalf of the United States under the antitrust laws to the effect that a defendant has violated said laws shall be *prima facie* evidence against such defendant in any action or proceeding brought by any other party against such defendant under said laws as to all matters respecting which said judgment or decree would be an estoppel as between the parties thereto. The United States shall file an antitrust proposal with the district court, publish in the Federal Register, and thereafter furnish to any person upon request, a competitive impact statement which shall recite: (1) the nature and purpose of the proceeding; (2) a description of the practices or events giving rise to the alleged violation of the antitrust laws; (3) an explanation of the proposal for a consent judgment, including an explanation of any unusual circumstances giving rise to such proposal or any provision contained therein, relief to be obtained thereby, and the anticipated effects on competition of such relief; (4) the remedies available to potential private plaintiffs damaged by the alleged violation in the event that such proposal for the consent judgment is entered in such proceeding; (5) a description of the procedures available for modification of such proposal; and (6) a description and evaluation of alternatives to such proposal actually considered by the United States.[1]

---

[1] U.S. Code. *http://caselaw.lp.findlaw.com/casecode/uscodes/15/chapters/1/sections/section_12.html.* 1999.

The other notable element of the U.S. Code that applies to the judgment process of alleged antitrust violations includes a public interest determination. The code states that before entering any consent judgment proposed by the United States under this section, the court shall determine that the entry of such judgment is in the public interest. For the purpose of such determination, the court may consider: (1) the competitive impact of such judgment, including termination of alleged violations, provisions for enforcement and modification, duration or relief sought, anticipated effects of alternative remedies actually considered, and any other considerations bearing upon the adequacy of such judgment; (2) the impact of entry of such judgment upon the public generally and individuals alleging specific injury from the violations set forth in the complaint including consideration of the public benefit, if any, to be derived from a determination of the issues at trial. While undergoing the procedure for a public interest determination the court may: (1) take testimony of Government officials or experts or such other expert witnesses, upon motion of any party or participant or upon its own motion, as the court may deem appropriate; (2) appoint a special master and such outside consultants or expert witnesses as the court may deem appropriate; and request and obtain the views, evaluations, or advice of any individual, group or agency of government with respect to any aspects of the proposed judgment or the effect of such judgment, in such manner as the court deems appropriate; (3) authorize full or limited participation in proceedings before the court by interested persons or agencies, including appearance amicus curiae, intervention as a party pursuant to the Federal Rules of Civil Procedure, examination of witnesses or documentary materials, or participation in any other manner and extent which serves the public interest as the court may deem appropriate; (4) review any comments including any objections filed with the United States under subsection (d) of this section concerning the proposed judgment and the responses of the United States to such comments and objections; and (5) take such other action in the public interest as the court may deem appropriate (Title 15, Chapter 1, Section XVI).

The U.S. Code is frequently referenced during antitrust proceedings, but the Sherman Antitrust Act, passed on July 2, 1890, is really the cornerstone of antitrust policy in the United States. The Sherman Act prohibits contracts, combinations, and conspiracies in restraint of trade in interstate commerce. Among the agreements prohibited by the Sherman Act are those that involve price fixing; allocation of markets or customers; and boycotts of competitors, suppliers, or customers. The Sherman Act also condemns monopolization. Specifically, the Sherman Act has two main provisions:

- *Sec. 1. Trusts, etc., in restraint of trade illegal; penalty.* Every contract, combination in the form of trust or otherwise, or conspiracy, in restraint of trade or commerce among the several States, or with foreign nations, is hereby declared to be illegal. Every person who shall make any such contract or engage in any such combination or conspiracy hereby declared to be illegal shall be deemed guilty of a felony, and, on conviction thereof, shall be punished by fine not exceeding $10,000,000 if a corporation, or, if any other person, $350,000, or by imprisonment not exceeding three years, or by both said punishments, in the discretion of the court.

- *Sec. 2. Monopolizing trade a felony; penalty.* Every person who shall monopolize, or attempt to monopolize, or combine or conspire with any other person or persons, to monopolize any part of the trade or commerce among the several States, or with foreign nations, shall be deemed guilty of a felony, and, on conviction thereof, shall be punished by fine not exceeding $10,000,000 if a corporation, or, if any other person, $350,000, or by imprisonment not exceeding three years, or by both said punishments in the discretion of the court.

It is not at all clear from the act precisely what was intended by the Sherman Act; some scholars maintain that it was no more than a requirement for the courts to create antitrust laws via legal precedent. Early court rulings by Justice Rufus Peckham led to what became known as the Peckham Rule. Peckham noted that competition tends to lower prices, increasing the volume of trade. Thus, combinations, contracts, or conspiracies intended to restrict output or raise prices are illegal, while combinations intended to increase output are legal.

There was significant opposition to the Sherman Act at the time of its early enforcement. Many, including President Teddy Roosevelt, felt that the act was quite vague, and preferred legislation that enumerated a list of proscribed activities. It is a fair criticism even today to say that it is very difficult to tell in advance whether an action complies with antitrust legislation. President Howard Taft, however, felt that enforcement of the Sherman Act was quite reasonable. Generally, the courts ruled that mergers and trusts intended to increase prices were illegal, and all others were legal.[1]

Randolph Preston, a University of Texas economics professor notes that starting with the breakup of Alcoa in 1945, and most recently in the Microsoft case, the courts employ a two-part test for the Sherman Act. First, does the firm have monopoly power in a product market? Absent monopoly power, a firm has no ability to reduce competition or influence prices. Second, did the firm use illegal tactics in acquiring or maintaining the market power? In particular, did the firm engage in "the willful acquisition or maintenance of that power as distinguished from growth or development as a consequence of a superior product, business acumen or historic accident?" That is, monopoly itself is not illegal, but some

---

[1] Kaserman, David and John Mayo. *Government and Business: The Economics of Antitrust and Regulation.* Ft. Worth, TX: Dryden Press. 1995, 65–74.

means of acquiring and preserving it are illegal.[1] Mr. Preston's questions will be addressed as allegations related to the Microsoft case are discussed.

The Clayton Act enacted in 1914 adds detail to the Sherman Act, supplementing and extending Sherman Act enforcement. The Clayton Act does not carry criminal penalties, but it does allow for monetary penalties three times the amount of damage created by the illegal behavior. The Clayton Act prohibits various kinds of business behavior that tends to lessen competition or monopolize trade. Among the activities prohibited by the Clayton Act are exclusive dealing arrangements, acquisitions, and mergers that tend to lessen competition.[2]

Section 2 of the Clayton Act prohibits price discrimination that would lessen competition. This law is not applicable in selling to final consumers, but usually it is applicable when selling to firms that compete with each other. The logic of the act is that a lower price of an input to one firm will create a competitive disadvantage for other firms. In addition, Section 2 prohibits a variety of subterfuges for concealing price discrimination, including brokerage commissions and indirect payments, and prohibits buyers from attempting to extract illegal discounts from sellers.

Section 3 of the Clayton Act prohibits a variety of exclusionary practices that lessen or eliminate competition. This section rules out a variety of transactions between the seller of a good and buyers who are themselves firms. The section has been interpreted to prohibit the following elements:

- Tying (must buy one good to get another)

- Requirements tying (buyer agrees to buy all its needs from the seller)

- Exclusive dealing (buyer agrees to deal with seller)

- Exclusive territories (buyer agrees to operate only in specified region)

- Resale price maintenance (buyer agrees to a minimum resale price)

- Predatory pricing (pricing below cost to eliminate a competitor)

In order to be illegal, these activities must reduce competition or work to exclude or eliminate a competitor.

[1.] McAfee, Preston. *Competitive Solutions*. Princeton University Press. 2002. p 204–210.

[2.] *http://www.swlaw.edu/programs/504.htm.* 2002.

The Role of the FTC and Other Antitrust Laws

In addition to the Sherman Act's broad prohibitions was another feature that disturbed some people: the lack of an agency specifically charged with implementing and enforcing the act. Prior to the Sherman Act, the Interstate Commerce Act provided for an administrative agency that would develop the expertise necessary to aid enforcement of the act. Thus, there was ample precedent for such an agency. Dissatisfaction with the performance of the Department of Justice in enforcing the Sherman Act fed growing sentiment for some form of interstate trade commission to supplement the department's efforts. Interestingly, support for this notion came both from those hostile to big business and from big business itself. As a result, the FTC Act was enacted on September 26, 1914 (Kaserman 1995, p 74). As a result, the Federal Trade Commission Act enacted in 1914 created the FTC and gave it a mandate to prevent unfair methods of competition a deceptive practices. The FTC commissioners have the ability to decide quickly that a practice is unfair and move to stop it. The FTC commissioners have a quasi-judicial standing in such proceedings and can issue restraining orders and stop deceptive practices. For example, the FTC has prevented Volvo from showing ads with monster trucks crushing other cars but not the Volvo, because the advertisements were staged. In particular, the roof supports of other cars were cut, while the Volvo was reinforced.

The FTC Act does not generally create more powerful antitrust enforcement than the Clayton Act. However, it was recently interpreted to prohibit suggesting collusion to competitors. Under either the Sherman or Clayton Act, the suggestion that firms might fix prices, if turned down, has not lessened competition and probably did not attempt to monopolize, and therefore is not illegal. The FTC Act, however, may make such a suggestion illegal. In fact, the major provision of the act empowers the FTC to discipline certain business behavior: "Unfair methods of competition in or affecting commerce, and unfair or deceptive acts or practices in or affecting commerce, are hereby declared unlawful." (Sec. 5(a)(1))

Joel Klein, Assistant Attorney General Antitrust Division notes that in terms of enforcement, there are three main ways in which the federal antitrust laws are enforced: criminal and civil enforcement actions brought by the Antitrust Division of the Department of Justice, civil enforcement actions brought by the Federal Trade Commission, and lawsuits brought by private parties asserting damage claims. The Department of Justice uses a number of tools in investigating and prosecuting criminal antitrust violations. Department of Justice attorneys often work with agents of the Federal Bureau of Investigation (FBI) or other investigative agencies to obtain evidence. In some cases, the Department may use court authorized searches of business, consensual monitoring of phone calls, and informants equipped with secret listening devices. The Department may grant immunity to individuals or corporations who provide timely information that is needed to prosecute antitrust violations, such as bid rigging or price fixing. A provision in the Clayton Act also permits private parties injured by an antitrust violation to sue in federal court for three times their actual damages plus court costs and attorneys' fees. State attorneys general may bring civil suits under the Clayton Act on behalf of injured consumers in their states, and groups of consumers often bring suits on their own. Such follow-on civil suits to criminal enforcement actions can be a very effective additional deterrent to criminal activity. Most states also have antitrust laws closely paralleling the federal antitrust laws. The state laws generally apply to violations that occur wholly in one state. These state laws are enforced similarly to federal laws through the office of state attorneys general.[1]

Of a number of amendments to the antitrust laws since 1914, three are particularly important: the Robinson-Patman Act, the Wheeler-Lea Act, and the Celler-Kefauver Act.

In 1936, Congress passed the Robinson-Patman Act, amending Section 2 of the Clayton Act. The Robinson-Patman Act was introduced for the avowed purpose of protecting small independent retailers and wholesalers from the growing inroads of chain stores. Indeed, its supporters described it as an "anti-chain store" bill. The bill, initially introduced in the House of Representatives and incorporated into the act as passed, was drafted by the counsel for the United States Wholesale Grocers Association.

One of the most basic changes made by the Robinson-Patman Act, in Section 2(a), broadened the test of illegality in Section 2 of the Clayton Act, so that it now reads "where the effect of such discrimination may be substantially to lessen competition or tend to create a monopoly in any line of commerce, or to injure, destroy, or prevent competition with any person who either grants or knowingly receives the benefit of such discrimination, or with customers of either of them." The addition, indicated by italics, has been interpreted as a deliberate effort to protect individual competitors rather than to promote the process of competition.

In 1938, the Wheeler-Lea Act amended Section 5 of the Federal Trade Commission Act by adding "unfair or deceptive practices in commerce" to the prohibition of unfair methods of competition. This amendment responded to a court decision that held that the FTC was not authorized to take action against a firm charged with false and misleading—indeed, medically dangerous—advertising of a thyroid obesity remedy when no effect on competition could be shown, but only deception of customers. The Wheeler-Lea Act also strengthened the administrative power of the FTC by providing that an order of the commission becomes final unless appealed to the courts within 60 days.

---

[1]   Klein, Joel, Assistant Attorney General Antitrust Division. Pamphlet: *Antitrust Enforcement and the Consumer,* US Department of Justice. Washington, D.C.: U.S. Government Printing Office, 1996.

The Celler-Kefauver Act of 1950 amended Section 7 of the Clayton Act, converting it from a prohibition against acquisition of part or all of the outstanding stock of a competitor into a general antimerger statute. The act made two major changes. First, the prohibition against the merger was made general by including asset acquisitions as well as those of share capital. Second, the test of illegality was broadened from a lessening of competition solely between the acquired and acquiring companies to a lessening of competition "in any line of commerce in any section of the country."[1]

Antitrust: IBM and AT&T

Franklin Fisher, professor of economics at MIT provides that the IBM case was perhaps the greatest antitrust fiasco of all time. Brought in January 1969 as the last act of the Johnson administration's Antitrust Division, it went to trial in 1975 and remained there until 1981. In the IBM case, the DOJ charged that the company used its market power and its control of standards to deter innovations by competitors and to extend its dominance into new market segments in which it did not necessarily offer the best projects.

In early 1982, the case was dismissed by stipulation, both sides agreeing that it had been "without merit." It spawned a large number of private suits, a few of which were "without merit." It also spawned a large number of private suits, a few of which were settled and all of which were won by IBM. I have had much to say about the reasons for the fiasco and, especially, about the bad economics of the Justice Department, and shall not linger over those subjects here except as they are relevant to the analysis of innovation and monopoly leveraging.

Among IBM's actions that were alleged to be anticompetitive were some of the bundling or tying variety. These fell into two main categories: the bundling of software or services with hardware, and the bundling of formerly separate hardware components together.

Until 1968, IBM offered systems support and software to purchasers or lessees of its computer systems at no separately stated charge. The government alleged that this was anticompetitive because it raised entry barriers into the computer systems market that IBM was said to have monopolized. As discussed earlier, monopoly leveraging can make sense as a device to protect against entry into the originally monopolized market, so such a claim requires one at least to know some facts before dismissing it.

Unfortunately for the claim, the facts are dispositive. The government's position was that bundling forced other hardware producers also to offer such a bundle

[1]   Baldwin, William. *Market Power, Competition, and Antitrust Policy.* Boston: Irwin Publications in Economics. 1987, 80–81, 92–99.

and that this made entry more difficult. This was quite untrue. In the first place, IBM' s bundling did not prevent the enormous growth of an independent software industry with very many participants, so that it would have been (and was) easy for hardware manufacturers to acquire the necessary software to produce a bundle. This made the supposed entry barrier merely a question of raising capital, and there were no capital barriers to entry.

Second, IBM's bundling in the relatively early days of the computer industry was a response to consumer demand. The bundle effectively provided insurance, a guarantee that computers would function and solve users' problems. That was highly desirable in a period in which computers were great, unfamiliar, frightening beasts. When a community of users arose that did not need the bundle, bundling diminished, starting in 1968.

The important point here is the following: had consumers not wanted the bundle, IBM's bundling would have made entry easier, not harder. Other hardware manufacturers could have offered hardware without the bundle to customers who wished to dispense with the latter. This would have made their products more attractive relative to IBM. Instead, most manufacturers offered their own bundle. It is noteworthy that, when IBM unbundled, Honeywell ran advertisements proclaiming that they still offered "the same old bundle of joy."

This was not a case of monopoly leveraging, IBM neither succeeded nor could reasonably have expected to succeed in monopolizing software and services. Moreover, as should now be plain to any reasonable person, IBM had no monopoly to leverage or protect.[1]

David Hart, Associate Professor at Harvard University, notes that under the Reagan administration DOJ ultimately abandoned the case, but not before IBM's market position had begun to erode in the face of new competition. Some of these competitors were Japanese firms whose questionable practices with respect to intellectual property might have been pursued more vigorously by the U.S. government had DOJ not been locked in conflict with IBM. New competitors closer to home offered mini- and microcomputers, products that IBM was reluctant to offer in part because they cannibalized its mainframe computers, but in part because it feared antitrust recrimination. Once IBM decided it had to get into the personal computer market, the company made a series of decisions that eventually ceded the bulk of the profits from this highly successful effort to Microsoft, Intel, and other firms. Whether antitrust concerns influenced these crucial choices is a matter of debate. What seems clear is that the antitrust case changed the terms of competition, contributing to the pursuit of a greater variety of technological paths than IBM would probably have pursued had it been left to its own devices.

[1.] Ellig, Jerry. *Dynamic Competition and Public Policy: Technology, Innovation, and Antitrust Issues*. Cambridge, MA: Cambridge University Press, 2001, 147–148.

U.S. v. AT&T, filed in November 1974, built on a number of precedents, too. The Truman-era DOJ had accused AT&T of illegally crushing its competition in telephone equipment manufacturing. With the support of DOD, AT&T settled that suit on favorable terms in 1956, maintaining its major technological assets, including Western Electric and Bell Labs. However, the consent decree did compel it to license its entire patent portfolio and to stay out of non-telephone markets. The foundational patents for the semiconductor industry were among those licensed; the strictures on AT&T competition in this area facilitated the growth of new firms that later became household names. The Federal Communications Commission (FCC) also shaped AT&T's business and technological environment in the 1950s, 1960s, and 1970s, permitting competitors to offer innovative products and services while limiting AT&T's responses. The most dogged of these competitors was MCI, which ultimately filed a private antitrust case against AT&T while pressing its advantage in the FCC.

In its 1974 case, DOJ reiterated its concern about AT&T's dominance of the equipment market, suggesting that competition would unleash a burst of technological innovation. The Modified Final Judgment in the case provided an opportunity to test that contention, because it forced AT&T to divest its local telephone operating companies and lifted the 1956 consent decree. The technological consequences of the AT&T case, like those of the IBM case, remain disputed. Some observers attribute the accelerated deployment of fiber optic lines, the development of the wireless industry, and even the growth of the Internet to the breakup, whereas others lament the downsizing of Bell Labs and the chaos in the management of the national communications system. Again, what seems clear in this case as well is that antitrust policy altered the spectrum of technological opportunities in an important sector by expanding the number of players and shifting the relationships among them.[1]

Box 1 lists key dates in the eventual demise of the Bell System.

The Case of Microsoft

Standard tradeoff analysis proceeds something like this. If Microsoft is permitted to integrate its Internet browser with the Windows operating system, there might be harms to competition as well as benefits to consumers. Microsoft could acquire market power in the browser market, but it could also offer consumers a superior product at a lower price. Integration should be prohibited if the harms associated with the extension of market power exceed the consumer

---

[1.] Hart, David. "Antitrust and Technological Innovation." *Issues in Science and Technology Online*. 1998. (5–7). Available at: *http://www.nap.edu/issues/15.2/hart.htm.*

Box 1: Bell System Key Dates

November 20, 1974	DOJ files antitrust suit charging anticompetitive behavior, and seeking breakup of Bell System.
February 4, 1975	AT&T formally denies all charges.
June 21, 1978	Case reassigned to Judge Harold Greene.
September 11, 1978	Judge Greene lays down new schedule for discovery and trial preparation.
November 1, 1978	DOJ files its first statement of contentions and proof, settling out detailed charges.
September 9, 1980	Judge Greene schedules beginning of trial for January 15, 1981.
January 15, 1981	Trial begins with opening arguments.
January 16, 1981	Judge Greene grants parties' request for recess until February 2, 1981 to work on a concrete, detailed proposal for settlement.
January 30, 1981	Judge Greene extends recess through March 2, 1981.
February 23, 1981	DOJ advises court it will not be able to approve a final agreement by deadline; settlement talks break off.
March 4, 1981	Trial resumes; testimony begins.
March 23, 1981	Defense Secretary Caspar Weinberger.
July 1, 1981	DOJ rests its case.
July 10, 1981	AT&T files motion for dismissal.
July 29, 1981	DOJ requests 11 month delay to permit Congress to consider amendments to S.898.
August 3, 1981	AT&T begins its defense.
August 6, 1981	DOJ says it will pursue case while Administration seeks passage of amended S.898.
August 10, 1981	DOJ says it would drop case if acceptable legislation enacted.
August 17, 1981	DOJ files reply to AT&T dismissal motion, saying it will pursue case.

Box 1: Bell System Key Dates (Continued)

September 11, 1981	Judge Greene rules on dismissal, dropping some charges, but permitting bulk of case to go forward.
October 26, 1981	Court sets schedule that will end AT&T testimony by January 20, 1982. Judge Greene indicates a verdict could be handed down by end of July, 1982.
December 31, 1981	DOJ announces that parties have resumed discussions to try to bring the case to a resolution.
January 8, 1982	Antitrust suit dropped after AT&T accepts government's proposal.
January 1, 1984	Bell System no longer exists.[1]

1. David Massey. AT&T Divestiture or "Breaking Up is Hard to Do!" Events that led up to the demise of the Bell system. 2002. *http://www.bellsystemmemorial.com/att_divestiture.html.*

benefits of integration. Integration should be permitted if the consumer benefits of integration outweigh the harms.

Dynamic analysis suggests a different set of cost and benefits, based on different paths that innovation might take. If Microsoft is permitted to integrate the browser with Windows, it has greater incentives to proceed with similar types of innovations in the future. Such integration may reinforce the position of Windows as the dominant computer operating system, and so firms that want to compete with Microsoft might have to develop an operating system that can displace Windows. If Microsoft is not permitted to integrate the browser with Windows, then browsers and other software applications will compete as free-standing products. Microsoft's competitors will have much stronger incentives to develop software applications on the Windows platform.

Clearly the direction of innovation is different under the two scenarios. The first scenario is more likely to produce competition among operating systems and suites of associated applications, whereas the second creates more competition among individual software applications within the Windows standard. Simply stating the tradeoff reveals the difficulty of determining the efficient answer. The answer requires not just an evaluation of the immediate consequences of Microsoft's business practices but also accurate predictions of alternative futures.

A rule-based approach lessens the burden on courts in individual cases. Perhaps the court need not assess whether consumers will be better off if Microsoft or its competitors receive superior incentives to innovate. All that is necessary is a more general understanding of typical results in cases of this type. If market dominance fostered by product bundling tends to produce superior innovations, then Microsoft should be left alone. If superior innovations come from markets where there is more competition in the production of individual components, then Microsoft should be prosecuted. Or perhaps bundled systems and unbundled, more open systems each tend to produce better results under different, objectively identifiable circumstances.

Although such an approach makes court decisions easier, it offers no shortcuts for academic researchers. If courts are to develop general rules, how are they to know which rules tend to produce the best results, if not by accessing research on past situations where different approaches were tried?

In a dynamic economy, antitrust policy implicitly involves a choice among alternative innovative paths. Implicitly or explicitly, enforcers are betting that their intervention will give consumers a better stream of price, product, and service improvements in the future.

Microsoft and the Current Business Environment

Details of Microsoft's ongoing legal happenings continue to become known to the public. For example, in recent news Microsoft was able to prevent the nine states litigating against it from introducing as evidence documents in which computer makers complain about the software company's tentative settlement with the Bush administration. The documents are critical to the states' case because they contain allegations that Microsoft is continuing to exercise unfair business practices, even after settling its case with the federal government (Zarate 2002, p. 1).[1]

In another more dramatic recent event, Bill Gates said on April 24, 2002 that a set of proposed antitrust penalties could force Microsoft to stop selling Windows. Gates' essential contention was the states' proposal, that would force Microsoft to continue licensing older versions of Windows, would lead to fragmentation of the operating system and consumer confusion. In effect, Gates noted that "Obsolete versions of Windows are a drag on the ecosystem," and that this would create "confusion in the PC consumer market" (McCullah 2002, p. 1).[2]

In one recent development, a Microsoft witness had the affect of aiding the government's case in an instance where a computer scientist testifying for Microsoft contradicted some of his earlier statements and acknowledged that a stripped-down version of the Windows operating system may be "technically feasible." One part of the states' 42-page proposed sanctions calls for a version of Windows with components such as Internet Explorer, Windows Media Player, and HTML help system deleted. In addition, the witness acknowledged that Windows XP embedded a "componentized" version of the operating system designed for

[1.]   Zarate, Robert. Cagey MS Moves to Seal Case. Wired Online. *http://www.wired.com/news/ antitrust/0,1551,52238,00.html*. Accessed 1 April 2002.

[2.]   McCullagh, DeClan. Gates: Complying "Not Feasible." Wired Online. *http://www.wired.com/ news/antitrust/0,1551,52060,00.html*. Accessed 1 April 2002.

use in specialized devices such as cash registers and automatic teller machines, which could be potentially configured to operate on a desktop and could satisfy the requirements of the states' proposal. Among all the legal challenges facing Microsoft, it would be in Microsoft's interest to carefully screen witnesses in order to reduce the likelihood of damaging their own cause (Zarate 2002, p. 1).[1]

Microsoft will likely face legal challenges for many years simply because of its undeniably large influence on the personal computing industry. Microsoft's continued desire to vigorously oppose any threats to competitive advantages that it believes were gained legitimately should serve the company well. The Justice Department, however, has shown that it too is a willing participant in legal battles where it believes that the public interest is best served by challenging Microsoft's business practices.

The department of Justice filed its complaint in the current litigation on May 18, 1998, although it followed from prior judicial proceedings. Indeed, nearly four years earlier, on July 1994, the Department had filed an earlier complaint. In the original action, the government focused on Microsoft' s use of exclusionary contracts that limited the ability of rivals to compete in the relevant market. A striking feature of that case is that it settled on the same date that the complaint was filed, with Microsoft's agreement to remove the offending contractual provisions. In the earlier proceedings, there was much debate as to whether Microsoft's exclusionary conduct actually had an impact on its market position. Microsoft's position rested originally on its selection by IBM to provide the operating system for its newly introduced personal computer. Since IBM's product then set the standard for personal computers, the Microsoft system MS-DOS, would be part of that standard. Microsoft's dominance in providing operating systems for personal computers was achieved by building on that advantage. In these circumstances, it is questionable as to whether its exclusionary policies were also an important factor and whether discontinuing these policies would have any remedial effect. Indeed, Microsoft's ready acceptance of the settlement terms suggests an answer to the latter question. Although the government charged that it was all the more important to remove artificial barriers when natural barriers were already present, it could also be the case that the artificial barriers imposed only nonbonding constraints. Following the agreed-upon settlement between Microsoft and the Department of Justice, the proceedings took a surprising turn. Under U.S. law, all proposed consent decrees must be approved by a federal court to confirm that the settlement serves the public interest. In this case, the court rejected the settlement and ruled it would not lead to a more competitive market outcome.

Two years passed, and Microsoft proceeded to introduce a new version of its operating system software that included Internet browser technology. The government cried "foul," and on October 20, 1997 returned to court asking that Microsoft

---

[1] Zarate, Robert. "It's Possible' Expert Hurts Microsoft." Wired Online. *http://www.wired.com/ news/antitrust/0,1551,52275,00.html* Accessed 1 April 2002.

be found in contempt for violating the earlier judgment. It argued that Internet browsers were a separate product, and that in distributing the two products together, Microsoft was requiring computer manufacturers "to license and distribute… [its] Internet browser as a condition of obtaining a license for… [its] Windows 95 operating system." The district court agreed on December 11, 1997, but again the court of appeals found for Microsoft on grounds that the United States presented no evidence suggesting that Windows 98 was not an "integrated product" and thus "exempt from the prohibitions" of the earlier judgment.

From this tangled background, the Department of Justice filed its current complaint. And while it included an allegation regarding the anticompetitive effects of a tying arrangement between Internet browsers and operating systems, the new action was far broader than that. Microsoft was now charged with using various methods to buttress its monopoly position, which taken together violated the monopolization provisions of the Sherman Act.

Chapter 9

The Digital Divide:
Policy Myth or Political Reality?

Jolene Kay Jesse

In September 2000, during the height of the U.S. presidential campaign, Republican candidate for President George W. Bush pledged $400 million to create 2,000 new community technology centers (CTCs) every year. The candidate touted the centers as examples of effective ways to provide Internet access, skills training, and computer literacy. Democratic candidate for President Al Gore quickly reiterated his own support for community technology centers, similarly pledging to build 2,000 centers by the year 2002 for those people who need it most.

By 2002, however, then-President George W. Bush proposed to cut all funding for the two most prominent federally funded CTC programs in his fiscal year 2003 budget: the Community Technology Centers program within the Department of Education and the Technology Opportunities Program within the Department of Commerce. Citing a new Commerce Department report, the Bush Administration justified the cut because of statistical evidence that the gap in access to computers and the Internet among various ethnic, racial, and geographical groups had decreased, and that more and more Americans were going on line at a speed that seemed to be outpacing the need for government intervention.

What happened to change President Bush's stance on federal investments in CTCs from that of Candidate Bush? Budget realities? Changes in philosophy? Priorities in other areas? A lessening of the Digital Divide? Probably some or all of those. Is there a Digital Divide? What is it? How do we measure it? How did it become a government policy priority? What are community technology centers and where did the impetus for them begin? What are the implications of addressing or not addressing the Digital Divide?

Awareness of the growing gap in computer and Internet use among certain groups in American society dates back to the 1980s. Before that, computer use by Americans was relatively low, not least because computers were huge and complicated to operate. With the rise of Apple and IBM personal computers and the Microsoft Corporation's easy to use operating systems and applications, more and more people discovered the benefits of computing without having to learn the language of computers. New software packages, instead, translated what humans wanted into a language understandable by machines at the push of a button (or the click of a mouse). And computers became small enough to fit into homes and offices without the need for constant technical oversight. They were still very expensive, however, so only a few households could afford them. Their use also required a certain level of technical literacy in order to work them properly.

The development of the Internet lagged behind the computer revolution, but as more and more Americans gained access through faster and faster modems, the potential of the new medium for advanced communication in commerce, education, and organization began to become apparent. The explosion of companies doing business solely by Internet communications (the dot-coms) reached a peak in 2000 before the Internet business bubble burst and many of the companies filed for bankruptcy.

The promises of the Internet (faster, cheaper, more universal commerce and communications), while a reality for many, are still anticipated for some. Are there those who are systematically left out of this revolution? Statistics do show that age, race, ethnicity, central city or rural residence, poverty, educational attainment, and gender have all served as filters that block access to the communications and income-generating potential of computers and the Internet. Are these filters insurmountable? In other words, are there ways to make the computer revolution more inclusive? Probably. The question is, how? Is more universal usage an eventuality that simply needs time for people to jump on the bandwagon? After all, haven't new technologies—televisions, telephones, CD players—ultimately become affordable and their use filtered down to all income groups. Or does the narrowing of gaps in information technology access require some kind of intervention, which brings the computers to the people, offering training in and access to a medium that is functional only when the equipment is available and people become computer literate?

These types of questions form the basis of policy-making regarding government investments in all kinds of goods and services. The government very often invests funds in projects that politicians feel are important. Government investment ensures that attention will be paid to an issue, and frequently leads to more private investment as companies and manufacturers often see government expenditures as a sign of future growth in a product, area, or service. Decisions involving the investment of public money are usually made when the market exhibits inefficiencies in meeting the needs of special segments of the population, or when such investment is perceived as able to expedite the development of new technologies, workforce segments, or research in areas of national priority. Determining when government investment in a priority area is beneficial, and when it dampens market efficiencies by limiting competition or imposing too many regulations, is a fine line that government policy makers must walk. In the case of the Digital Divide, statistics seemed to indicate a market inefficiency in meeting the needs of a sizable portion of the public. How to meet those needs is a big question, and often the policy eventually adopted depends on who gets to interpret the facts and, subsequently, whose voice is heard loudest in the debate.

The Clinton Administration did adopt significant policies and programs to bridge what it saw as a huge market inefficiency—the Digital Divide. These included instituting several CTC programs, including the TOP program in the Department of Commerce and the Community Technology Centers program in the Department of Education. In addition, Clinton began a multi-billion–dollar program to help schools, school districts, and public libraries adopt computers, telecommunications infrastructure, and Internet technology. The Bush administration has subsequently proposed the elimination of the CTC programs, but has left intact the telecommunications portion of Clinton's Digital Divide policy. Currently, we find that the Digital Divide is rapidly changing. Is this because of or in spite of government efforts to mitigate it?

Defining the Divide

In 1994, the Clinton Administration's National Telecommunications and Information Administration (NTIA) within the Department of Commerce released statistical data on Americans' computer ownership and use. The NTIA survey, entitled *Falling Through the Net, A Survey of the 'Have Nots' in Rural and Urban America,* for the first time presented U.S. census data that revealed patterns of telephone, computer, and modem usage in the United States. The data showed that various groups were being left out of the dynamic information technology sector, although Internet access was not an issue in this first report because its use was not yet widespread.

The NTIA data illustrated that Americans were most likely to own a computer if they lived in an urban area but not the central city; made at least $35,000 a year; were Asian American or White; and had some college education. College education nearly doubled the likelihood of computer ownership, with approximately 30 percent of households with a college-educated individual owning a computer, while approximately 15 percent of households with only a high school education owned a computer. The number dropped to 6 percent for those with less than a high school education. These numbers also varied by rural–urban residence, with non-central city urban dwellers more likely to own a computer and central city residents falling the furthest behind. Older Americans (those over 55) were also less likely to own a computer in 1994 (only around 12 percent did).

The numbers that most interested policy advocates and policy makers, however, seemed to be the gulf between the haves and the have-nots along racial and ethnic lines. Nearly 40 percent of Asian-American households and 30 percent of White households owned computers in 1994. The number was around 10 percent for Black households and approximately 12 percent for Hispanics. Only 6.4 percent of Black households in rural areas had a computer in the home. While the division between the haves and the have-nots was a gulf between suburbia and central cities and rural areas, between high income and low income, between older and younger Americans, and between the educated and the less educated, the divide along ethnic lines propelled the notions of technology haves and have-nots into the discourse of civil rights. The corresponding rhetoric was almost hyperbolic, with, for example, NAACP President Kweisi Mfume calling the gap "technological segregation." Such language was sure to catch the attention of the public and policy makers.

The term "Digital Divide" appeared some time after 1994, although no one is quite sure who first used it. It was an immediately powerful catchphrase, however, that succinctly and effectively captured the sense of the monumental gulf that seemed to separate the technology haves and have-nots. It was easy to say and easy to conceptualize. Its meaning was also readily expandable from computer ownership, to access to the Internet, to broadband access, to ownership of broadcast and Internet companies. In other words, it is a term that is easily adapted.

By 1998 the NTIA was using it in its follow-up report, *Falling through the Net: Defining the Digital Divide*. The report asserted that, although computer usage was growing among all groups, the gap between the haves and have-nots, especially along racial and ethnic lines, was becoming wider than ever. While Whites and Asians increasingly had computers and Internet access in their homes, American Indians/Eskimos/Aleuts, Blacks, and Hispanics relied heavily on public libraries and community centers for access, if they went online at all.

By this time, the rhetoric of the Digital Divide had already changed from one of economic opportunity to a matter of civil rights. As notions of "digital government" began to surface, access to the Internet began to be seen as imperative for full participation in American politics, economics, and society. President Clinton made Internet access a key public policy objective as he announced his multi-billion–dollar initiative that would wire every classroom and library in the country by the year 2000 and every home by 2007. The Internet was perceived as the ultimate educational, economic, social, and democratic forum, and so access was a right that should not be exclusive to those who could pay for it.

This rhetoric made the Digital Divide debate potentially more contentious. Groups like the Leadership Conference on Civil Rights, the Civil Rights Forum on

Communications Policy, the Loka Institute, OMB Watch, the Center for Civic Networking, and NetAction began to advocate for Internet access in low-income communities as a tool for political and economic empowerment. Government funding for CTCs began to grow within the Departments of Education and Commerce, and a special tax was placed on telephone service to fund Clinton's computer initiative for low-income schools (the e-rate). While there was some debate over the effectiveness of these programs, for the most part Digital Divide initiatives thrived during the Clinton years. Things would change, however, with the election of George Bush to the presidency in 2000.

Power (PCs, Hardware, and Software) to the People[1]: The Evolution of a Government Policy

Almost as soon as the first computers were put into K–12 classrooms in the 1970s and 1980s, there was talk of technology "haves and have-nots." Computers were instantly seen as a tool of the future, one that our children would need to know how to use to become economically successful in the America of the rapidly approaching twenty-first century. But, obviously the wealthiest schools and school districts could afford computers for their students, while the poorer school districts lagged behind. Schools are physical entities and school districts are spatial entities with strict geographic boundaries. These easy delineations serve to make the gaps between schools and school districts both measurable and traceable. That school districts are also often divided by income, geographical location (rural-suburban-urban), and, despite efforts at integration, by ethnicity and race, the groups who are the "have-nots" are also easily identifiable. Wealthy school districts are, for the most part, suburban and White. Poorer school districts are in the inner city or out in rural communities and serve a disproportionate population of children who are Black, Hispanic, Native American, immigrants, from low-income families, and children with disabilities.

Efforts to bridge the computer gap began in the non-profit education sector. The first CTC was opened in 1983 in Harlem by a retired schoolteacher, Antonia Stone, who opened a center in the basement of a housing development in Harlem funded by private donations. The CTC movement was relatively quiet throughout the 1980s, with private funds financing centers that made up a loose network under the umbrella organization of Stone's Playing-To-Win non-profit organization. In the 1990s, rapid advances in computer usage and software applications intensified the talk of technology haves and have-nots, and increasingly became

---

[1]. Advertising slogan for a Washington, D.C.-based computer company in the Fall 2002.

part of the public policy debate. CTCs also gained a foothold in securing federal funding from the National Science Foundation (NSF).

How did the CTC movement, then, evolve from a loose network of education-oriented computer resource centers in a handful of inner cities to a collection of various interest groups advocating everything from free Internet access to more minority-owned broadcast companies? Numbers are powerful tools. They can tell a story that is compelling and can lend a sense of urgency to a cause. In the case of the Digital Divide, the numbers were so lopsided that the divide seemed insurmountable. The Digital Divide also had the advantage of touching on two important government policies with a long history of government investment and inclusion in the civil rights discourse—public education and universal service.

Although we now take federal funding for and regulation of public education for granted, it is important to remember that there wasn't a U.S. Department of Education before 1979. In fact, no cabinet-level department existed to deal with education issues before 1953 when President Eisenhower created the Department of Health, Education and Welfare (HEW). There was no direct federal funding for public education until 1958, and that was extremely limited and focused on math and science in response to the perceived national crisis in U.S. science and math education posed by the Soviet Union's launch of the Sputnik satellite.

Instead, for most of our nation's 200-plus years of existence, education was a purview of each state, with funding gleaned from state and local taxes, and priorities set by state education offices. The problem with the state-centered system was that in some states different groups were given more resources than others. In many states, funding for education still comes from local property taxes, creating a situation in which some school districts with a higher tax base have more resources than poorer districts. In addition, throughout the nineteenth century and into the twentieth century, some states, especially some southern states, created "separate but equal" school systems in which Whites and Blacks (and in some cases Hispanics and American Indians) were segregated, and in which minority school districts were given substantially fewer resources. In 1954, the Supreme Court ended segregation with their landmark ruling in *Brown v. Board of Education of Topeka, Kansas*. This ruling by a federal court inevitably tied federal education policy to civil rights.

Funding for public education became a significant federal government priority in the 1960s during the height of the Civil Rights movement and President Johnson's War on Poverty. Johnson launched the most extensive effort to reach low-income citizens in the United States since the New Deal initiatives of President Roosevelt in the 1930s. In 1965, Congress passed the Elementary and Secondary Education Act (ESEA), which for the first time created a significant funding category for schools with a high proportion of students whose family incomes were below the poverty level. By 1968, twelve cents of every dollar spent on education in the United States came from the federal government. Currently, a little less than ten cents of every dollar spent on education comes from federal sources.

By prioritizing federal education spending on low-income schools and school districts, the ESEA made explicit the division between schools and school districts that were haves and have-nots. As the Presidency and Congress have changed from Republican to Democratic control and back again, the debate over education vacillates between giving the states more control (the Republican position) and using federal funding and requirements to regulate education at the federal level to ensure equal access to educational resources (the Democratic position). It is not surprising, then, that a Democratic President Clinton would choose to initiate funding programs to equalize access to the educational opportunities that computers and the Internet promised.

Equally important to government policy making regarding the Digital Divide during the 1990s was a history of telecommunications policy in the United States that emphasizes universal service. The 1994, NTIA report began with the statement: "At the core of U.S. telecommunications policy is the goal of universal service—the idea that all Americans should have access to affordable telephone service." This was the first time that the government had coupled the notions of universal telephone service and computer access in a government report.

Universal service is not a new idea. The push for universal telephone service has been a government policy since the Communications Act of 1934. Universal service subsidies also have historically funded electricity and natural gas usage in rural and poor areas. What was new in Falling through the Net was the assertion that universal service was "an individual's pathway to the riches of the Information Age, [and that] a personal computer and modem are rapidly becoming the keys to the vault." Such statements further reflected President Clinton's own brand of populism, which promoted access to education and information as the pathways to success for all Americans. In other words, if you could get plugged in, you could realize the American Dream.

The backdrop of the universal service mandate helped to further the Digital Divide policy discourse. In 1996, Congress expanded the goal of the universal service program to cover information technology, requiring "reasonable comparability" of services and rates among rural, urban, and suburban areas, paid for through the Universal Service Fund financed through telephone service charges.

Federal Investments to Bridge the Digital Divide

After the release of the 1994 NTIA report, the Clinton Administration began an escalation of new initiatives to bridge the Digital Divide. The non-profit sector also used the NTIA report to lobby Congress for more federal funds for CTCs. One result was the Technology Opportunity Program (TOP) launched by the NTIA in 1994. To date, TOP has awarded 555 grants to organizations in all 50 states, the District of Columbia, Puerto Rico, and the U.S. Virgin Islands, totaling $204.9 million in federal funds, leveraged by $282 million in local matching funds. TOP funding is available to non-profit organizations, including local and state governments, hospitals, fire departments, universities, and other educational and public service organizations, to set up programs that deliver digital network technologies to serve the public interest. TOP projects include initiatives that address the areas of lifelong learning, community and economic development, government and public services, safety, health, and culture and the arts. TOP's mandate is to support innovative models that can be replicated by other communities.

The Community Technology Centers program in the Department of Education (ED) began in 1999 and has distributed more than $150 million to 337 projects since its inception. School districts, universities, community colleges, state and local governments, and other non-profit organizations run the projects. Centers provide access to information technology as well as teach technology skills and literacy to children and adults. They include pre-school and family activities, after-school and youth programs, adult education, elder services, career development and job preparation courses, job training, and other services. The program is designed to fund model programs demonstrating the educational effectiveness of technology and provides access to people who do not have computers in their homes, especially in central city and rural areas and economically distressed communities.

Both TOP and ED's Community Technology Centers are highly competitive programs that offer seed money to get centers started in underserved communities, or funds to further the work of already existing CTCs. Most awards average around $500,000. Funding is usually for three years or less and is contingent upon matching funds from private and other government sources. In fact, all TOP and CTC grantees that received initial public monies, raised from 102 to 150 percent more in private funding for their initiatives. Receiving a federal grant, in other words, provided centers a base of resources to begin operations and a sense of legitimacy with which they could then go to other sources to fundraise.

As federal money for CTCs became increasingly available, their numbers grew exponentially. The Playing-to-Win Network of Antonia Stone was an alliance of six CTCs in 1990. By 1995 they had 52 affiliates, and that number grew to 200 by 1997. In the summer of 2002 the number of affiliated organizations in the CTCNet Consortium (the successor umbrella organization to Playing-to-Win) was 700. Today, they claim 1,000 members in all 50 states, the District of Columbia, Puerto Rico, and the Virgin Islands. In addition to the CTCNet Consortium, there is the America Connects Consortium for CTCs and the Digital Divide Network, all umbrella organizations that collect information, disseminate how-to manuals and other publications, and advocate for CTCs throughout the United States. Other organizations that are affiliated with CTCNet which have national constituencies include groups that promote programs for people with disabilities (Alliance for Technology Access), for Hispanics (ASPIRA Association, Inc.), for low-income Americans (Morino Institute, Technology for All), for African Americans (the

National Urban League), for union employees (Service Employees International), as well as other groups.

Parallel to this investment in CTCs, the Telecommunications Act of 1996 levied the e-rate tax (e for education) on long-distance telephone service with the proceeds being used to subsidize computers and Internet access purchases for classrooms and libraries in poor communities. The e-rate has generated about $2.25 billion dollars per year since 1998 for school and library technology subsidies, and represents the federal government's most wide-reaching technology program. Schools and school districts are allocated an amount of funding determined by the number of students qualifying for the national school lunch program (a measure of the poverty of the population), which they can then use to purchase telecom infrastructure, computers and Internet access on the open market. Subsidies range from 20 to 90 percent. In a study conducted by Goolsbee and Guryan[1] (2002), they found that the e-rate subsidy in California successfully increased Internet investments among urban schools with large African-American and Hispanic student populations, although it was less successful in rural areas. While Goolsbee and Guryan emphasize that California schools were already investing at an increasing rate in Internet connections and computers before the e-rate program, they estimate that the program accelerated computer and Internet access in urban areas by about four years.

As lawmakers continued to ponder the Digital Divide numbers, the economy was in full swing, powered by the growth of Internet companies that needed more and more technologically savvy workers to fill high-tech jobs. "Techies" without college degrees were landing jobs at Internet start-up companies with high salaries and benefits, relaxed working environments, and stock options that considerably increased their net worth. At the same time, non-technology companies were becoming more and more wired, and the government, at all levels, was realizing the benefits of computer technology for efficient administration.

The technology explosion created jobs faster than the American education system could produce tech-literate Americans to fill them, and so the "shortages" in the high-tech labor force also entered into public policy debates. While many wanted the United States to allow more immigration of high-tech workers from overseas, there was also a movement to include members of underrepresented groups in education and training that would allow them to more fully participate in the high-tech workforce. Diversity in the high-tech workforce was extolled as the best way to grow the economy and increase the production of technology products and services that would be designed to meet the needs of all people.

In 1998, Congress set up the Commission on the Advancement of Women and Minorities in Science, Engineering and Technology Development (CAWMSET),

---

[1]   Goolsbee, Austan and Jonathan Guryan. 2002. *The Impact of Internet Subsidies in Public Schools*. NBER Working Paper #9090: *www.nber.org/papers/w9090*.

also known as the Morella Commission after Constance Morella (a Republican Congresswoman from Maryland). The Commission was made up of industry representatives and academics appointed by a bipartisan group of lawmakers, with a staff of NSF employees and an interagency steering committee with representatives from all the cabinet-level departments and agencies in government.

The Commission's September 2000 report, *Land of Plenty: Diversity as America's Competitive Edge in Science, Engineering and Technology,* used statistics to back up an assertive agenda for diversifying the nation's high-tech workforce. This agenda included strengthening standards-based science and technology education at the K–12 education level; "aggressive" government interventions to encourage more women and minorities to enter technology fields at the post-secondary level; a national campaign to change the image of science and technology careers; a movement to hold employers accountable for developing mechanisms to advance the career development of women, minorities, and people with disabilities; and the establishment of an ongoing government panel to oversee progress on these issues.

At the end of the Clinton Administration, there seemed, therefore, to be bipartisan support for the President's policies to wire the country, to promote technology education, and to subsidize Internet resources. Government revenue surpluses at the end of the 1990s and early 2000 also fueled the optimism surrounding the adoption of additional government programs to close the Digital Divide. In October 2000 the NTIA released the fourth report in their series on the Digital Divide entitled: *Falling Through the Net, Toward Digital Inclusion.* This edition documented the increasing use of the Internet across all groups and a somewhat decreasing Digital Divide. However, the report concluded, there were still significant gaps, and the government's priority should remain to provide access to computers and the Internet for all groups as equitably as possible.

So, in the fall of the year 2000, notions of diversity and closing the Digital Divide were part of a political landscape that was awash in technology-driven plenty. Both Democrats and Republicans supported measures to pump new resources into education. Al Gore, the Democratic Presidential candidate, despite a rocky political campaign, seemed poised to ride on the coattails of a popular President and a booming economy right into the White House. His Republican opponent, George W. Bush, was also using education as a major component of his Presidential campaign. CTCs and other technology programs seemed to fit into everyone's agenda. The only source of debate appeared to be how much money would go to them and how many centers would be created and supported.

From the Digital Divide to a Nation Online

The Clinton Administration's last NTIA report, *Falling Through the Net: Toward Digital Inclusion*, stressed that Black and Hispanic households still lagged behind their White and Asian-American counterparts in computer ownership and home Internet access. Approximately 51 percent of all American households had a computer in August 2000, compared to 42.1 percent in 1998 and 24 percent in 1994. The percentage of African-American households with a computer had grown to 32.6 percent in 2000, compared to 23.2 percent in 1998 and 10 percent in 1994. For Hispanic-American households the percentage that owned a computer in 2000 reached 33.7 percent, compared to 25.5 percent in 1998 and 12 percent in 1994. So, in 2000, there were definitely signs of digital expansion across racial and ethnic groups, but as the report emphasized, it was still uncertain if there was real inclusion.

The gaps among racial and ethnic groups persisted at almost all income and education levels. In fact the gaps in computer ownership did not disappear between White and African-American households, until the $75,000 a year income level. The gap between White and Hispanic households remained over 10 percentage points even at the $75,000 income level (87 percent of White households that made more than $75,000 a year owned computers while only 76.1 percent of Hispanics at that income level did so).

Falling Through the Net did document rapid changes in Internet access. In 1998 only 26.2 percent of American households had Internet access. By 2000, the percentage had increased to 41.5 percent. Internet penetration was still highest, though, among middle- and upper-income households (those that made $35,000 or more) and households with at least one member that had some college education. A member of the family with a bachelor's degree more than doubled the likelihood that a household would be wired to the Internet compared to households with only a high school diploma as the highest educational attainment (66 percent versus 29.9 percent). A high school diploma also more than doubled the likelihood of Internet penetration compared to households without a high school diploma (29.9 percent versus 14.8 percent).

In 1998, only 11.2 percent of Black households had Internet access. Twenty months later, in 2000, that number had doubled to 23.5 percent. Similarly, 12.6 percent of Hispanic households were online in 1998, but 23.6 percent were wired in 2000. At the same time, 56.8 percent of Asian-American households and 46.1 percent of White households had Internet access in 2000. While these documented increases in Internet access among Blacks and Hispanics were important, the report intoned that the gap between minority access and the rest of America was actually widening, mostly because Whites and Asian Americans were adopting home Internet use at a much more rapid rate. So, while in 1998 the gap between African-American household Internet access and the national average was 15 percentage points (11.2 percent and 26.2 percent respectively), in August 2000, the gap was 18 percentage points (23.5 percent versus 41.5 percent). Even as Black households were gaining ground, the report indicated, they were still falling behind. A similar widening of the gap was also evident for Hispanics. Falling furthest behind were Black and Hispanic households that made less than $15,000. Only 6.4 percent and 5.2 percent of those households had Internet access.

Besides household access, *Falling Through the Net* also examined individual American's Internet use (as opposed to household Internet access) at home and at public access points (e.g., at work, schools, libraries, and other venues). It found that 55.7 percent of Americans did not use the Internet at all in 2000. Of those who were using the Internet, 35.7 percent had home access, while 19.4 percent used the Internet outside of their homes, and 44.4 percent used the Internet both at home and in outside locations. But these numbers were very different when broken out by race and ethnicity. Figure 9–1 shows that Whites and Asian Americans were more than twice as likely to be using the Internet at home, or both at home and outside. Hispanics and African Americans were much more likely to not be on line at all. Those who did use the Internet were more likely to access it outside the home than their White and Asian counterparts. The implications of the report were that out-of-home access was inferior to having access in the unhurried atmosphere of one's home.

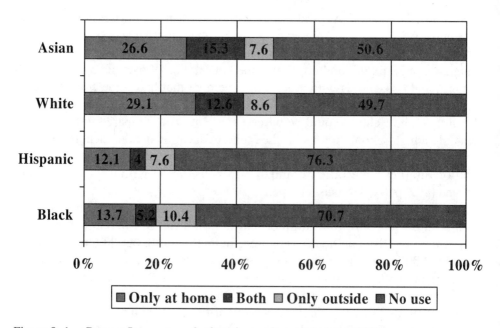

Figure 9–1   Percent Internet use by location and race ethnicity, 2000.
(Source: NTIA and ESA, U.S. Dept. of Commerce, using U.S. Bureau of the Census Current Population Survey Supplements.)

The report concluded by reiterating the Clinton Administration's emphasis on universal access by intoning that "Until everyone has access to new technology tools, we must continue to take steps to expand access to these information resources." This included government investment in computers for schools and libraries as well as CTCs.

Sixteen months after the last Clinton Administration report, the new Bush Administration released its first Commerce Department report on computer and Internet use. Entitled *A Nation Online: How Americans are Expanding Their Use of the Internet*, the report presented a different picture of the adoption of computers and Internet usage among the various groups in the United States, and nowhere in the report was the term Digital Divide even used. Based on a Census survey conducted in September 2001, *A Nation Online* reported that 56.5 percent of American households surveyed had a personal computer and that 88.1 percent of those homes subscribed in some way or other to the Internet. Thus, almost half of the U.S. households surveyed were connected to the Internet. Compared to the 18.6 percent who had Internet access in 1998, this represented a huge increase in home Internet use over a three-year period. At an individual level, the survey found that over half of the people in the United States (approximately 143 million or 53 percent) were using the Internet, up from 44.5 percent in 2000.

Unlike the previous reports, *A Nation Online* focused its analysis more on individuals and their computer and Internet use rather than households. Individuals, the survey found, were increasing their use of the Internet outside the home, especially at work and at school. Internet use was also bridging across the traditional divides of income and education, as well as race and ethnicity. While the report did acknowledge that some people were more likely to be Internet users than others, over time, *A Nation Online* suggested, Internet use was becoming more equitable.

The division along racial and ethnic lines still persisted in the 2001 report. While 71.2 percent of Asian Americans and 70 percent of Whites used computers, only 55.7 percent of Blacks and 48.8 percent of Hispanic Americans did so. Internet use among Whites and Asian Americans was around 60 percent, while its use among Blacks and Hispanics were 39.8 percent and 31.6 percent respectively. The report pointed out, however, that the rate of growth of Internet use among Blacks and Hispanics was higher than that for Whites and Asians. (See Figure 9–2.)

For the first time, *A Nation Online* also addressed the question of subgroups within the standard racial/ethnic categories. Especially among the Hispanic population, Internet use was highly affected by Spanish language use. Among those Hispanics in which Spanish was the only language spoken, only 14.1 percent were online, in contrast to 37.6 percent of Hispanics in households where Spanish was not the only language spoken.

Most dramatically, *A Nation Online* documented the phenomenal rise in computer and Internet use among the nation's children. Children and teenagers were the most likely to be computer users and the most likely to be on the Internet.

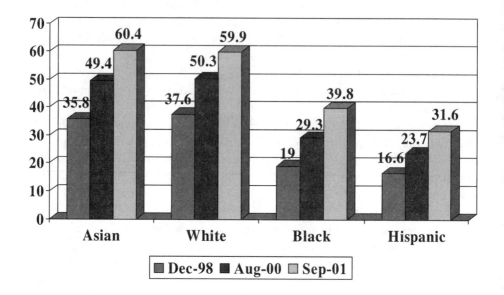

Figure 9–2   Internet use by race/ethnicity.

(Source: NTIA and ESA, U.S. Department of Commerce, using U.S. Census Bureau Current Population Survey Supplements)

92 percent of children aged 9–17 were computer users and 68.6 percent of children in that age range used the Internet. Children were followed closely by adults in the 20–50 age range, those most likely to be in the workforce and using computers and the Internet on the job. The report credited the presence of computers in schools for decreasing the gap in computer and Internet use among children across income and other demographic groups. Having children also increased the likelihood of a household both owning a computer and using the Internet.

There were still documented gaps among children of different races and ethnicities. Black and Hispanic children were between three to four times as likely to have access to computers only at school, not at home. 38.9 percent of Hispanic and 44.7 percent of Black students had access to computers only at school, while 70.7 percent of Asian-American students and 71.8 percent of White students had access to computers at both home and school. Black and Hispanic students were also twice as likely to not be using a computer at all. (See Figure 9–3.)

A Nation Online advanced the notion that there was a significant reduction in inequalities in Americans' use of computers and the Internet. Its conclusions focused on who was online, and the report found that more and more Americans were able to gain access to technology with less regard to income, educational attainment, and racial/ethnic differences, although some inequalities still remained. While not all Americans, and not all American school children could afford to have computers and home Internet access, schools and libraries were

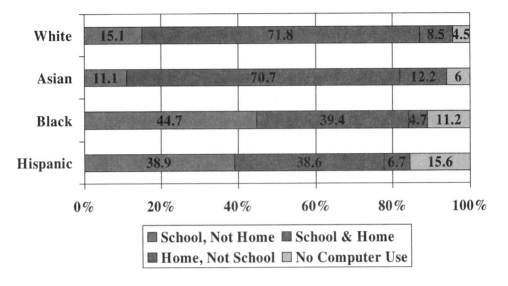

Figure 9–3   Computer use among 10–17 year-olds by race/ethnicity and location, 2001.
(Source: NTIA and ESA, U.S. Department of Commerce, using U.S. Census Bureau Current Population Survey Supplements)

touted as the great equalizers to computer and Internet use. Declining computer prices, the report suggested, would also lead to a more equitable distribution of computer ownership and Internet access.

The timing of the report (February 2002) coincided with the release of President Bush's fiscal year 2003 budget proposal. In it, the President's priorities for education were explicitly laid out. Resources were redirected to implementing the No Child Left Behind Act of 2001 (NCLB), which mandated testing of all school children in grades 3 through 8 and more funding for poorer school districts. At the same time, the Bush Administration's budget advocated consolidating special programs, especially individual technology access and education programs, into more flexible block grants that states could use to invest in priority areas and thereby reduce duplicate or unsuccessful programs.

On the chopping block were many special programs with strong advocate bases, including the Community Technology Centers program in ED and TOP in the Department of Commerce. Federal funds for the programs were already halved in the fiscal year 2002 budget appropriations bill. The Bush Administration cited *A Nation Online* as proof that the programs were no longer needed, and that market forces, rather than government interventions, would be more effective in closing the remaining Digital Divide. This assertion set off a maelstrom of counter assertions by CTC advocates who quickly produced their own reinterpretations of the Commerce Department data and stepped up their efforts to advocate for technology access programs. In February 2003, Congress refused to eliminate the programs, and with a bipartisan coalition of lawmakers, appropriated funds for both programs through fiscal year 2003. Surprisingly, Bush left the significantly larger, though less visible e-rate program completely intact.

The Benton Foundation, administrator of the America Connects Consortium of CTCs, was one of the first organizations to release a reinterpretation of the Commerce Department data in a policy brief dated March 2002. The foundation found that the Digital Divide was widening along educational, income, racial, and geographic lines, rather than lessening as *A Nation Online* asserted. Focusing on household data, the Benton Foundation brief emphasized that the gap in Internet access between the wealthiest and the poorest American households was growing rather than receding. The continued lag in African-American and Hispanic-American households' Internet access was of continuing concern for the Foundation. Also cited was the gulf in broadband Internet access between rural and urban areas, which the authors of the brief found to be widening despite upward trends in both areas. The brief concluded that, "community technology investments are paying off," and cited the Community Technology Centers Program and TOP as prime examples of successful government investment in technology access and education.

In July 2002, the Leadership Conference on Civil Rights Education Fund and the Benton Foundation released a jointly commissioned report exploring in depth the inequalities apparent to them in the data from *A Nation Online*. Entitled *Bringing a Nation Online: the Importance of Federal Leadership*, the report focused attention on the segments of society that were lagging in computer and Internet use. Troublesome for these organizations was that "Significant divides still exist between high and low income households, among different racial groups, between northern and southern states, and rural and urban households. For people in these communities, the enormous social, civic, educational and economic opportunities offered by rapid advances in information technology remain out of reach." The stakes in the Digital Divide, according to the report, were too high to be left for the vagaries of the marketplace. Government intervention had assured and continues to assure that Americans have equal access to telephone service. It had also made the footholds that disadvantaged groups had made in bridging the Digital Divide a reality. Information technology and its promises, according to the authors, require government intervention to open the doors of opportunity to all segments of American society.

Bringing a Nation Online asserts that the real message of *A Nation Online* is that "federal leadership matters." Without federal leadership, computer availability and Internet access in almost every school and library in the nation would not have been a realizable goal—a significant factor, according to the Commerce Department report, for achieving high technology usage among America's young people. Such federal investment is also needed, according to *Bringing a Nation Online*, to effectively train America's domestic workforce for the rapidly advancing challenges in technological literacy required for the workplace of the twenty-first century. Access is not enough, the report intoned. There needs to be effective programs to train people to successfully use technology for their own economic and political empowerment, as well as increased efforts to provide appropriate Internet content for all groups in society. The collaborative nature of public and private partnerships exemplified by the TOP and the Community Technology Centers Programs, the report emphasized, had contributed to the amazing advances in computer and Internet use by Americans in less than a decade.

Doubting the Divide

Throughout the Clinton Administration there were always critics of the Digital Divide. Conservative analysts from the Heritage Foundation and the American Enterprise Institute (AEI), along with libertarians from the Cato Institute denounced the Clinton Administration for using politics to interfere with the market. These critics found the Clinton policies unnecessary at best, and at worst, some half-jokingly accused the Clinton Administration of pandering to high-tech executives in Silicon Valley and elsewhere looking for government handouts in exchange for large contributions to the Democratic Party.

What the Commerce Department data really chronicled, these critics asserted, were the normal gaps in new technology adoption. As evidenced by radio and television, economically disadvantaged groups lag behind those with resources in adopting new technologies. Eventually, as the price drops and technology becomes easier to operate, more segments of the population are able to afford and use it. This is the "trickle down" theory of technology adoption that has been observed for most high-tech gadgets, including VCRs and cellular telephones, which some minority groups have taken up at even faster rates than majority groups. Indeed, analysts from the Cato Institute point out that computers and the Internet have spread faster within the American population than cars, televisions, and VCRs did.

Some of these critics have also suggested that Americans do not need computers and Internet access to lead productive lives. The Internet is more of an entertainment medium, rather than a skill promoting or empowering mechanism, and as such, will not necessarily promote better education, increased social interaction, job skill training, or any of the other wonders that the Clinton Administration had claimed it would. Subsidizing its use, therefore, would be akin to subsidizing cable television or movie tickets, an egregious waste of taxpayer's money that would be better spent on improved education and teacher training. And far from promoting social interactions or political empowerment, increased Internet use has been cited as alienating, decreasing social contact, and traditional community and family activities among ardent users. Moreover, most Americans who want to be online will be, these critics contend, and Internet penetration will never reach 100 percent simply because not everyone wants it.

This last point seems substantiated at least in part by a new survey by the Pew Internet and American Life Project. In *The Ever-Shifting Internet Population: a New Look at Internet Access and the Digital Divide*, the Project found that 42 percent of Americans are not currently online and that most don't feel they ever will use the Internet. The three top reasons that Internet non-users cited as the main impediments to going online were cost, lack of time, and "the Internet is too complicated and hard to understand."

Additionally, 52 percent of non-users claimed that they simply don't want or need to use the Internet. About 20 percent of non-users live in households that are connected to the Internet, and some see their non-use as a source of pride. Approximately 17 percent of self-identified non-users had once been users, a cohort that the report labels as "Net Dropouts." The Project found that only 24 percent of Americans are truly non-Internet users who have no access whatsoever or experience with the Internet.

Trickle-down economic critics also contend that government interventions such as CTCs may be too little too late. Rapid advances in technology coupled with rapid adoption by individuals could mean that the government builds a host of these centers only to find that no one wants to use them, and that the equipment all too soon becomes obsolete. In that way, the government has spent millions of taxpayer dollars and has not provided the kinds of services that would best meet the needs of the target populations.

Similarly, they see the e-rate program as deceptive, outrageously expensive and fraught with fraud and scandal. Derisively dubbing the program the "Gore Tax," for then-Vice President Al Gore who promoted the program, the AEI lambasted the Administration for trying to pressure telephone companies to hide the tax on consumers' long distance telephone bills by not making it a separate line item. The amount of money involved in the program ($2.25 billion a year), moreover, injected huge sums of money into the market so quickly that schools and school districts were inundated with ads for technology products and services, making it almost impossible for them to adequately assess their needs and make informed choices. Besides, AEI contended, the incentives weren't there for schools to make informed choices, but, rather, to spend the money quickly while it was available. This kind of resource glut could only produce scandals, and, indeed, Republicans in the House of Representatives in March 2003 stepped up a congressional investigation of the program claiming $200 million, if not substantially more, have been misappropriated since 1998.

Since the release of *A Nation Online*, other critics of Digital Divide rhetoric have also emerged, questioning both the reality of the Digital Divide numbers used by advocates, as well as the efficacy of relying solely on data for policy decision-making. Some have gone so far as to suggest that the focus of the Digital Divide rhetoric on the lag in technology adoption among minority groups comes dangerously close to creating negative stereotypes of minorities as technophobes or as charity cases. Such rhetoric can be patronizing and marginalizing in its own right, many scholars contend, focusing attention on hardware gaps rather than on the contributions that minority group members have already made to computer and Internet development and content. In other words, the Digital Divide concentrates on the negatives rather than the positives.

The problem with the Digital Divide rhetoric, according to some critics, is its single-minded focus on the gaps in technology access. The fear is that putting the spotlight solely on gaps reduces the likelihood of private investment in the areas and among the groups that seem to be slower to adopt technology. Many are asking for a more nuanced approach that goes beyond the data to acknowledge the contributions that African Americans and Hispanic Americans have made to technoculture—for example hip-hop and DJ music. There is a broader social context within which empowerment occurs, according to these critiques, that is being ignored. By only focusing on access, the argument goes, the interrelation between technology gaps and other social problems cannot be adequately addressed. It's not just about owning a computer and being able to log on to the Internet. It's about what people do once they're online, whether they can navigate the technology for their own advancement by becoming technologically literate, and finding Internet content that is socially, politically, economically, and culturally relevant.

In addition to those who advocate for a more positive and "holistic" approach to the divide and the "trickle down" economists, there are those who are trying to gain a more dynamic picture of just who is and who is not online. Using independent survey data, these studies attempt to disaggregate groups and use more sophisticated statistical analyses to uncover patterns that go beyond stereotypical groupings. Most of these find that some combination of income, age, education, and openness to technology are greater predictors of computer and Internet adoption than race and ethnicity. They also assert that the Digital Divide is lessening among various groups at greater and greater rates.

The Pew study cited above, for example, focused almost exclusively on non-Internet users. While the study did find African Americans and Hispanics to be less wired than their White counterparts, the report emphasized age and income/employment status as being greater predictors of Internet use and non-use. In addition, the study found that of the 40 percent of non-users who said they thought they'd go online some day, Blacks and Hispanics were more likely to be in this group than Whites.

A survey conducted by a Baltimore, Maryland, market-research company in the spring of 2003 found that there is no Digital Divide among their sample of 500 college-bound high school seniors. In fact, their survey found that African Americans were more likely to use computers in their college search than their White counterparts. Another survey of Hispanic Internet users completed by comScore Networks in January 2003 of 50,000 Hispanic Internet users found that Hispanics were online about the same amount of time as other Internet users. Hispanics on the Internet are, however, somewhat younger than the general online population, with 60 percent under the age of 34 compared to 50 percent of the total online population. The company also found that 51 percent of Hispanics who were online preferred English as their language of choice in the home.

While the online population seems to be in flux and diversifying rapidly, there is little consensus on how that was accomplished. Did government investment in wiring schools and libraries accelerate the use of computers and the Internet among underserved populations, thus making it more likely that they adopt technology in their homes? Did such investment give different segments of the American workforce job skills that would increase their participation in technology jobs? Or did Americans simply flock to the Internet as market forces reduced the price of technology?

While Internet use has grown, it is still troubling for some that there remains a lack of interest on the part of some segments of the population in computer science and information technology education and a subsequent lack of diversity in the U.S. information technology (IT) workforce. Despite the increased use of computers and the Internet by all segments of the population, there are still wide gaps by gender and race/ethnicity in the IT workforce. In fact, a recent study by the Information Technology Association of America (ITAA) found that there has been very little progress in increasing the numbers of women and minorities in the lucrative IT employment sectors from 1996 to 2000. During this time period, the percentage of women and some minorities in the IT workforce actually declined. This finding reinforces concerns that the Digital Divide is much larger than just Internet access, and that it could separate population segments in ways that would leave many behind as the whole structure of the American economy changes.

The Digital Divide in IT Education and Employment

Computer Science and IT education at the postsecondary levels has displayed a complicated trend in light of the explosion of IT jobs throughout the 1990s. From 1985 to 1998, the number of bachelors' degrees awarded in Computer Science fell by 38 percent (from 39,927 to 24,912). The rate of decline for women over this time period was 50 percent. During roughly the same time period (1983–1998), jobs in computer science, computer engineering, and systems analysis increased from fewer than 300,000 to over 1.5 million. Although it is conceivable that many young people during the late 1980s and 1990s opted to skip the bachelor's degree and enter the IT workforce without this credential, given the percentage of women in the IT workforce, it is unlikely that many women did so.

Since 1998, degree production in Computer Science at the baccalaureate level has increased, but has yet to return to the 1985 level, and there is some indication that since 2001, this upward trend has again gone down as the decline in IT employment options has made it a less desirable major. By gender, the rate of growth in bachelors' degrees has been much greater for men than for women. (See Figure 9–4) The rate of growth for underrepresented minorities (African Americans, Hispanic Americans, and Native Americans) exhibited only slow gains in absolute numbers, but some losses when looked at percentage wise (for example, African Americans earned 9.5 percent of the bachelors' degrees in computer science in 1997, but only 9.1 percent in 2000, although in absolute numbers they increased the number of degrees earned).

These data seem to reflect a lack of adequate computer and technology training for women and minorities at the elementary and secondary school level, which may discourage them from pursuing IT bachelors' degrees. Moreover, when representatives from these groups do pursue bachelors' degrees in IT or computer science, they are more likely to drop out before attaining their degree. Subsequently, the numbers of women and minorities who pursue graduate degrees in IT fields is even lower, as at every juncture they drop out of the science and technology education pipeline.

Dropping out of the IT education pipeline means that, for women and minorities, entrance into the IT workforce is a less likely career option. The 2003 ITAA report on IT workforce diversity cited above follows on the heels of a number of similar reports that have appeared since 1999, including the CAWMSET report. These reports reflect a growing anxiety that the nation can not keep up with the demand for IT workers, and especially during the heady years of the Internet startups in the late 1990s and the year 2000, warned of impending worker shortages. These reports opened up the discussion about possible solutions to workforce shortages. While importing foreign workers tended to dominate the policy discourse in Congress and the Executive branches, these reports highlighted the lack of diversity in the IT workforce pool, which relies heavily on white males and does a poor job attracting women and minorities.

Should we be concerned about this reliance on one segment of our population for our IT workforce needs? The Bureau of Labor Statistics predicts that eight of the fastest growing occupations and five of the largest growing occupations in the United States from 2000–2010 will be computer related. Predictions of worker shortages have been exaggerated in the past, but what is clear is that, according to the U.S. Census Bureau, the segment of the population we rely on most for IT workers (i.e., white males) is declining as an overall proportion of our population and our workforce in general. While white males make up about 35 percent of our population overall, they dominate the science and technology workforce at about 63 percent. And this population is aging rapidly. In 1993, 48 percent of science and technology workers in the United States were over the age of 40. By 1999 the percentage of those over 40 had increased to 56 percent.

The tragedy of September 11, 2001, moreover, made immigration on temporary visas a poor policy choice to deal with IT worker shortages. Currently, the wait time to obtain visas to immigrate, even temporarily to the United States for

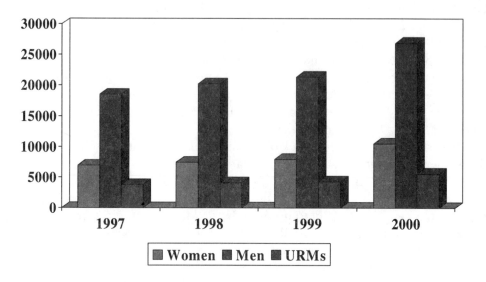

Figure 9–4   Bachelors' degrees awarded in computer science by gender, 1997–2000.
(Source: NTIA and ESA, U.S. Department of Commerce, using U.S. Census Bureau Current Population Survey Supplements)

work or study, has increased phenomenally as the Immigration and Naturalization Service (INS) deals with reorganizations and incorporation of some of it's duties into the new Homeland Security Agency. Individual interview requirements and other bureaucratic hoops will only extend the process, making importation of labor a much stickier business than in the past.

So, relying on our own domestic workforce to address worker shortfalls seems more and more an imperative, and may require greater outreach to groups currently underrepresented in the U.S. IT workforce. While our IT workforce may not be quite as old as the rest of the S&T workforce, it is still predominately white male. As Figure 9–5 shows, while the numbers of those employed in computer systems analysis and computer scientist occupations have increased at an ever-growing rate over the last 20 years, women have not kept pace with men in the field, and in 2001 experienced a small decline, even as the number of men had increased, possibly indicating that women may have been hit harder by the down market in IT employment.

The 2002 data on the participation of African Americans and Hispanic Americans in the mathematical and computer science workforce reveal that 7.3 percent and 5.1 percent of the workforce respectively came from those groups. This is despite that combined they make up almost 25 percent of our total population, and 21 percent of our total workforce.

What can be done about the apparent trend away from choosing IT fields on the part of women and some minority groups? What are the consequences of their exclusion from these fields? The authors of the various reports on diversity in the

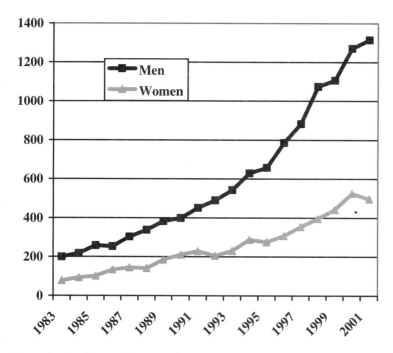

Figure 9–5   Men and women in computer system analysis and computer scientist occupations 1983–2001 (in thousands).

(Source: Data derived from the Bureau of Labor Statistics, Current Population Survey (unpublished tabulations)}

IT workforce contend that excluding any members of our population from our IT workforce jeopardizes the global leadership role of the United States, excludes groups from important avenues for citizen participation, and shuts them out from lucrative jobs that could lead to greater social mobility.

The question, however, is how to include these groups effectively. Is computer access enough? The Goolsbee and Guryan study found that access to computers had no appreciable impact on student test scores. Experience with computers or access to the Internet, furthermore, does not guarantee knowledge of the underlying principles of computer science or IT, nor does it necessarily prepare someone for an IT career. At the same time, never having seen or worked with a computer diminishes the likelihood that someone will choose IT or computer science as a college major or as a career.

Even more problematic for policy making in this area is the current downtrend in IT hiring. At the same time that ITAA released their IT diversity study, they also reported that hiring in IT fields will likely remain stagnant in the coming year. While less people are being laid off from tech jobs than in 2002, there has not been much growth in tech jobs in 2003, leaving those who were laid off last year still without employment options. Even more troubling, technology companies seem to be more and more willing to export tech jobs in order to take advantage of lower-cost workers overseas. This down market in IT employment provides little inducement to women and minorities to choose IT, and renders programs that encourage them to do so on problematic footing if there are no jobs for these groups after their expectations have been raised.

Where Should the Government Invest?

During his 2003 State of the Union Address, President Bush proposed spending $1.2 billion for research on hydrogen powered automobiles. President Bush obviously thinks that hydrogen-powered cars are the wave of the future, and that government investment is a necessary catalyst for their development. Is this a good place for government to invest taxpayers' money? Does it merit public funds? Or should the development of hydrogen-powered vehicles be left to the private sector? We already have hybrid cars on the market, and the rising price of gasoline has led some consumers to look for more efficient alternatives. Shouldn't we leave the development of hydrogen-powered vehicles to market forces and instead invest our dwindling government resources elsewhere? But where? Education? Health care? Tax cuts? Military spending? How do we decide?

The Clinton Administration saw a real need to get America connected. The economy at that time relied heavily on new technologies, requiring an increasingly technically savvy workforce, and it was booming. The booming economy led to government budget surpluses that were spent on a variety of programs, including a little under $500 million for community technology centers. In addition, the Universal Service Fund administered by the Federal Communication Commission spent $2.25 billion per year since 1998 to provide discounts to schools, libraries, and rural health care providers, enabling them to access the Internet more easily. Were these good investments? How does one measure a good investment?

The Digital Divide, by many measures, is decreasing. More Americans are online than ever before, and an increasing number are buying computers and accessing the Internet from their homes, schools and workplaces. Is this despite of or because of government spending on information technology infrastructure and programs to bridge the Digital Divide? Or is it both? The case of the Digital Divide shows clearly how individual opinion often determines the interpretation of data. At times, the data appears to be uncontroversial—for example, there clearly was a wide gap among racial and ethnic groups in computer ownership and use in 1994. But when the data changes rapidly, when the differences become less distinct, policy options become less cut and dry and policy debates become more contentious. Do we still have a gap in 2001? Is the gap insurmountable? Or have things changed enough that we can feel confident that Americans of all age, income, education, gender, and racial and ethnic make-ups will continue to adopt technology at an increasing rate? Will they know what to do with it once it's in their homes? Will it lead to more advancement or to employment opportunities that will increase social mobility and engagement? Will government investment help or hurt the situation? These are all questions that policy makers must weigh before continuing or discontinuing government programs that address national IT needs.

Further Reading

NTIA Reports: The *Falling Through the Net* series can be found at: *http://www.ntia.doc.gov/ntiahome/digitaldivide/index.html.*

A Nation Online can be found at: *http://www.ntia.doc.gov/ntiahome/dn/index.html.*

CTC Links: CTCNet Consortium: *http://www.ctcnet.org/index.html.*

America Connects Consortium for CTCs: *http://www.americaconnects.net/.*

Digital Divide Network

http://www.digitaldividenetwork.org/content/sections/index.cfm.

OMB Watch: *www.ombwatch.org.*

Benton Foundation Policy Brief on the Digital Divide: *http://www.benton.org/publibrary/policybriefs/brief01.pdf.*

"Bringing a Nation Online: The Importance of Federal Leadership": *http://www.benton.org/publibrary/nationonline/bringing_a_nation.pdf.*

Community Technology Centers program in the Department of Education: *http://www.ed.gov/offices/OVAE/AdultEd/CTC/index.html.*

Technology Opportunities Program in the Commerce Department: *http://www.ntia.doc.gov/otiahome/top/.*

Doubting the Divide

Compaine, Benjamin M., *The Digital Divide: Facing a Crisis or Creating a Myth.* Cambridge, MA: MIT Press, 2001.

Young, Jeffrey R., "Does 'Digital Divide' Rhetoric do more Harm than Good?" *Chronicle of Higher Education*, November 9, 2001. *http://chronicle.com/weekly/v48/i11/11a05101.htm.*

Singleton, Solveig and Mast, Lucas, "How Does the Empty Glass Fill? A Modern Philosophy of the Digital Divide." *EDUCAUSE Review*, November/December 2000. *http://www.educause.edu/pub/er/erm00/articles006/erm0062.pdf.*

Thierer, Adam D., "A 'Digital Divide' or a Digital Deluge of Opportunity?" The Heritage Foundation Executive Memorandum, No. 646, February 1, 2000. And "How Free Computers are Filling the Digital Divide." The Heritage Foundation Backgrounder, No. 1361, April 20, 2000. *www.heritage.org.*

Demuth, Christopher, "The Strange Case of the E-Rate." American Enterprise Institute for Public Policy Research, On the Issues, July 1, 1998. And Glassman, James K. "Gore's Internet Fiasco." *American Enterprise Institute for Public Policy Research*, On the Issues, June 2, 1998. *www.aei.org.*

Lenhart, Amanda, *The Ever-Shifting Internet Population: A New Look at Internet Access and the Digital Divide*. Pew Internet and American Life Project, April 16, 2003. *www.pewinternet.org.*

"Technology Update: No Digital Divide," Student Poll, Volume 4, Issue 4, April 17, 2003, Art & Science Group LLC. *www.artsci.com.*

"Hispanic Internet Users in U.S. Now Exceed the Total Online Population of Many Major Spanish-Speaking Nations, comScore Reports," The Multicultural Advantage, *http://www.multiculturaladvantage.com/contentmgt/anmviewer.asp?a=604&z=2&isasp=.*

IT Workforce Reports and Statistics

Congressional Commission on the Advancement of Women and Minorities in Science, Engineering and Technology Development, September 2000, *Land of Plenty: Diversity as America's Competitive Edge in Science, Engineering and Technology. www.nsf.gov/od/cawmset.*

Report of the ITAA Blue Ribbon Panel on IT Diversity presented at the National IT Workforce Convocation, May 5, 2003, Arlington, VA. *www.itaa.org/workforce.*

Commission on Professionals in Science and Technology, July 2002, *Professional Women and Minorities: A Total Human Resources Data Compendium. www.cpst.org.*

Chapter 10

Information Technology Workforce

William Aspray

"You know something's up when Jack Kemp, Spence Abraham and Bill Gates find themselves on the same side of the barricades as Al Gore, Bob Graham and the folks at the National Council of La Raza," says the Wall Street Journal Editorial Page this morning (June 12, 2000). You sure do: you know that the Anti-American Axis of professional ethnics, cheap-labor hogs, campaign-contribution whores, low-IQ libertarian loonies, neocon nasties, fossilized Republican publicists and New World Order nogoodniks that has paralyzed the immigration debate is about to launch its long-awaited Ardennes Offensive in Washington, aimed at enacting a vast expansion of H1-B "temporary" indentured-servant visas under cover of the election."

—Peter Brimelow (*http://www.vdare.com/pb/matloff_h1b.htm*)

Help Wanted

The late 1990s was a remarkable time for information technology. A long bull market led companies across every sector of the U.S. economy to increase, improve, or replace their information systems. Firms invested in Internet technologies to make their supply chains more efficient and to serve their customers better. The dot-com boom underway in Silicon Valley was being replicated in a dozen other places around the United States, creating new products and services based on communication and information technologies. Meanwhile, some companies that had waited too long before acting or could not afford to change to new computer systems spent thousands or, often, millions of dollars in fixing Y2K software bugs in their "legacy" mainframe systems.

This boom in information technology created extraordinary business opportunities and tremendous wealth for individuals, organizations, and the national economy, but the good times were threatened by a shortage of qualified workers to drive this economic engine. In 1997 the largest trade association representing information technology, the Information Technology Association of America (ITAA), published a report entitled *Help Wanted: The IT Workforce Gap at the Dawn of a New Century*, indicating a shortage of 190,000 information technology workers in large and mid-sized U.S. companies. The following year the association published a second report, *Help Wanted 1998: A Call for Collaborative Action for the New Millennium*, based on a larger sample of companies, indicating a shortage of 346,000 workers. This represented 10 percent of the IT workforce.

Having heard from ITAA and others, the U.S. Department of Commerce prepared its own report, *America's New Deficit: The Shortage of Information Technology Workers*, late in 1997. Commerce's findings mirrored the first ITAA report, presenting an image of a serious and growing shortage of IT workers in the United States. The Commerce report, however, did something that no trade association report could do. It legitimated the issue, giving public acknowledgement of the workforce problem and recognition that it was an issue worthy of federal attention.

Not everybody was convinced of the arguments made by the ITAA and the Commerce Department. Bruce Barnow, a well-known labor economist from Johns Hopkins University, and some colleagues at the Urban Institute criticized the methodology of the first ITAA study. The General Accounting Office in a document released in March 1998 raised concerns about the research behind the Commerce Department report.

One concern had to do with the demand statistics. The two ITAA studies were based on a study of vacancy rates. In the first ITAA study, conducted for ITAA by the libertarian think tank, the Cato Institute, 2000 companies of 500 or more employees were surveyed. However, only 271 companies—a mere 14 percent—responded. With such a low response rate and small sample size, it was hard to tell if the linear projection to the entire U.S. workforce was at all accurate. For example, perhaps companies with high vacancy rates were more motivated to respond to the Cato Institute survey than those companies that were able to fill their positions, resulting in an over-representation of vacancies in the sample. The second ITAA study, of 1500 companies with 100 or more workers, had a better, but still low 36 percent response rate.

Another concern had to do with the supply statistics. As a measure of supply, the first ITAA study used bachelor degrees awarded in computer science in the United States. The National Center for Education Statistics—a reliable source—indicated that about 25,000 of these degrees were being awarded annually. ITAA contrasted this number with a projection by the Bureau of Labor Statistics that 95,000 new IT jobs would be created each year for the coming decade. ITAA further noted a national decline of 40 percent in the number of Bachelor's degrees awarded in computer science nationally since the high point in 1986. Two criticisms were raised of this argument. Even by the time of the first ITAA report there were strong indications of a major increase in national production rates for bachelor's degrees in computer science. Over the course of the late 1990s, bachelor's degree enrollments in computer science more than doubled in the United States. This recent increase in enrollments had been brought to the attention of the ITAA, and they were upbraided at least once publicly when they chose to ignore these inconvenient facts in a public presentation.

Even more troubling was the metric itself. It was clear that people enter into IT work from a multiplicity of career paths: learning on the job; enrolling in short courses offered by a variety of non-profit and for-profit educational and training organizations; matriculating in graduate programs in computer science although holding a bachelor's degree in a different subject; studying for a minor or certificate in computer science; enrolling in another kind of computing-related discipline such as computer engineering, management information studies, or information science; or gaining sufficient knowledge of information technology through degree programs in mathematics, the sciences, or business studies. Indeed, further analysis revealed that only about one-quarter of IT workers earned a bachelor's degree in computer science.

Historical Background

Although the political debate of the late 1990s was the most public and most rancorous political discussion of the IT workforce, it was not the first time that the supply of IT workers had been discussed as a national policy issue. The first conference on training personnel for the computing machine field, to use the parlance of the day was held at Wayne University in Detroit in 1954. The first stored-program computers in the United States had been placed in operation only several years earlier. Although there was a rapid growth in business uses of computers later in the decade, at this time the computer was still conceived of as a giant calculator, mostly for scientific and engineering applications. In this context, the personnel required for operating computers were mathematicians, often with doctoral degrees, who could write mathematical applications for the machines.

It was noted at the conference that there were only about 2,000 Ph.D. mathematicians in the United States in 1951. Leon Cohen from the National Science Foundation and F. J. Weyl from the Office of Naval Research expressed concern about how small this number was and how the shortage of mathematicians might harm growth in the construction and use of computing machines. Colonel C.R. Gregg of the Air Materiel Command referred to the wide interest in computing in Congress, the Office of the Secretary of Defense, and each of the military departments. In fact, during the 1950s, various mission agencies, including the Army, Navy, Air Force, and Atomic Energy Commission, sponsored a limited number of university research projects and provided for graduate student support and computer infrastructure. But there is no indication in the 1950s that the federal government took any actions to address a national computing labor shortage.

In the late 1950s and early 1960s, computer labor issues were addressed by the federal government primarily as part of the build-up of science and engineering in response to the Cold War. After the Sputnik crisis of 1957 and the passage of the National Defense Education Act, the NSF became more heavily involved in all areas of science education. Computing was not a special priority, but computer education was advanced through more general NSF grant programs for science curriculum development, teacher training institutes, and scholarships and traineeships for graduate students. NSF initiated a computer facilities program in the late 1950s, which, by the time of its termination in the early 1970s, had provided computers to approximately 200 U.S. colleges and universities for research and education purposes. The Advanced Research Projects Agency (ARPA, later DARPA), formed within the Department of Defense only four years earlier, began to support computer research in 1962. Within one year, it was providing more support to universities for computing research than all other federal agencies combined. DARPA funding helped to build up many of the top-ranked computer science departments, while NSF support helped to build up the academic computer science community more widely. Thus DARPA and NSF both supported, in an important but somewhat indirect way, training of the academic and industrial research leaders in the field.

Up through the early 1960s, the federal government had provided a patchwork of support, through a variety of programs in various agencies, for computing research and some computer education; but the federal government had really not addressed computing as a national concern up until that time. This attitude changed in the mid-1960s. Computing emerged as a concern of the federal government not only because of its growing importance to science and technology and to national defense, but also to education and the general welfare of the citizenry. The first six studies on the nation's computing needs and on the role of the federal government in meeting them were carried out during this decade.

The first of these studies, published in 1966 under the auspices of the National Academy, studied Digital Computer Needs in Universities and Colleges, was directed by J. Barkley Rosser, a mathematician at Cornell University. Unfortunately, the study was seen as too self-serving of the academic community, and Congress was unwilling to provide the hundreds of millions of dollars called for to build up the nation's teaching and research capacity in computing.

More effective was the report published in 1967 from a Panel on Computers in Higher Education, chaired by Bell Labs electrical engineer John Pierce and convened by the President's Science Advisory Committee. The influence of this report was seen in President Lyndon Johnson's 1967 message to Congress on Health and Education, which directed the NSF to set up its first office on computers and begin an intensive set of research projects to improve the nation's public education system through the use of computerized educational tools. One plank of President Johnson's Great Society program was to train computer scientists who could improve public education. This office also took responsibility for programs to enhance computer facilities and provide research grants to computer scientists.

Federal support for computing diminished in the late 1960s, at first to pay for the Viet Nam War and then through the Nixon Administration's unwillingness to support academic computing facilities. In 1970, with a perceived glut of mathematicians and physicists, the Office of Management and Budget slashed the NSF budget for fellowships and traineeships. Since fellowships for computer science were supported through the program for mathematicians and physical scientists, support was cut for computer science graduate students just as science policy leaders called for a major expansion of academic programs in computer science. A National Academy of Engineering conference on Computer Science Education in 1969 had projected the need for 1,000 new Ph.D.s per year by 1975. This represented a ten-fold increase over the current national output. Yet between 1968 and 1975 the NSF budget for computing dropped by 15 percent in real dollars; and calls for help from academic departments for new faculty and support for graduate students went largely unheeded.

By the late 1970s, there was a widely perceived crisis in academic computing. National production of doctorates had reached a peak of 244 in 1976—far short of the 1,000 called for by the National Academy. Universities could not find enough faculty, and a typical department might have a quarter of its positions vacant. Teaching loads were heavy. Funding for research and equipment was low, and as a result faculty research increasingly turned to theoretical projects that required little more than pencil and paper. However, this kind of research was not of primary interest to industrial employers who wanted to hire graduate-trained scientists and engineers with knowledge of the experimental side of computing. NSF convened a workshop in 1978, which published a report known by the name of its principal editor, computer scientist Jerome Feldman from the University of Rochester.

The Feldman Report recommended building strength in experimental computer science in the universities and providing the infrastructure to do so. In particular, the report recommended and the NSF funded about 25 centers of excellence intended to enable research universities to become strong in computer research and graduate education—in much the same way that a very few chosen universities had become strong through the sustained support of DARPA. This Coordinated Experimental Research Program was controversial. Limited funding meant that NSF had to redirect funding intended for individual researchers at many different universities into a much smaller number of large grants at only a few schools—to support the faculty, students, staff, and equipment in a research center. By the end of the 1980s, about 30 schools had received these grants, and a separate program was implemented to support primarily minority-serving institutions. With hindsight, the CER program, together with cooperation and financial investment from some of the leading industrial research laboratories into the universities, were seen as successful at building up a much broader base of research and education in computing across U.S. universities.

Much as the 1960s had seen the first national recognition of computing, the 1980s led to a major increase in national programs and budgets. In 1986 NSF established the Directorate for Computer and Information Science and Engineering, giving computing the same organizational stature as the major traditional sciences, such as physics, mathematics, chemistry, and biology. That same year, the National Research Council, the research arm of the National Academies of Science and Engineering, formed the Computer Science and Telecommunications Board to provide guidance to the federal government on public policy and technical issues related to computing and telecommunications.

Federal support for computing prior to 1980 had been focused primarily on enhancing national defense, beefing up the universities to provide research and education, or on using the computer as a tool of social welfare in the Johnson Great Society. The Reagan Administration instead supported computing because they saw it as a means to build national competitiveness, especially against a Japanese economic threat. The computer was envisioned as spawning an industry that could drive national economic well-being as well as supporting the growth of other national industries.

One result was the building of national supercomputer scientists in five locations, which scientists around the country could access through ARPANET and other networks. It was as much a program to train computational scientists as it was a program to provide research facilities. This led in 1987 to a proposal by the Executive Branch's Office of Science and Technology Policy for a major five-year investment in high-performance computing and networking, with an annual investment across federal agencies of approximately $500 million for computer science research and development. This proposal was promoted in Congress by Senator Albert Gore in 1989 and implemented during the early years of the Clinton Administration as the High Performance Computing and Communication

Initiative. The budget eventually grew to exceed $1 billion and involve twelve federal agencies. Computing had made it big time; and this helped to increase the number of computer scientists and computer engineers the universities could train. The initiative was continued later in the 1990s under a successor program known as the Next Generation Internet.

Only with the establishment of the H1-B program in 1990 and the revisiting of the H1-B quotas in the late 1990s did national policy concerning IT workers move beyond the interests of NSF, DARPA, a few other science-oriented agencies, and the National Academies to broader and more mainstream political issues such as immigration and taxation. This was in part because the personal computer and the Internet had become major drivers of the economy and had brought IT beyond the defense and scientific communities, and large business, into every walk of life. With this historical context in mind, it is now possible to return to the workforce debates of the late 1990s.

A National Problem

Despite the limitations in the arguments put forward by the ITAA and the Commerce Department, an apparent shortage of IT workers had come to be perceived as a national problem. By 1998, local, state, and especially federal politicians were feeling pressure from industry to do something about the shortage of IT workers. In the spring of 1998, Congress held hearings on the needs of US companies for IT workers. The White House was encouraging the NSF to study the issue. That same year, the Foundation made an award to Computing Research Association to undertake a study of this issue, in collaboration with five other professional computing societies (American Association of Artificial Intelligence, Association for Computing Machinery, Computer Society of the Institute of Electrical and Electronics Engineers, Society of Industrial and Applied Mathematics, and USENIX Association).

CRA published its findings in 1999—in a book-length report from a study group led by Peter Freeman of Georgia Institute of Technology, who later became the assistant director of NSF responsible for managing the Foundation's computing directorate. *The Supply of Information Technology Workers in the United States* was one of the first studies on this subject to appear, and it received wide attention in the Washington policy community. It reviewed the political context, helped to define who is an IT worker, described the various sources of IT workers, evaluated whether there was a shortage of IT workers, identified serious problems with the data available for making policy decisions, discussed some of the contextual issues that limited the opportunities for companies, universities, and the federal government to act effectively, considered special workforce issues related to women and under-represented minorities, and highlighted concerns about the ability to train the next generation of IT workers because of the migration of faculty and graduate students from the academic to the industrial sector.

An outcome of congressional hearings in the spring of 1998 was the American Competitiveness and Workforce Improvement Act of 1998. While the law is mainly known for increasing the annual cap on the temporary, non-immigrant (H1-B) visas that could be awarded to specialty workers in high technology, it also provided support for a study on high-technology labor market needs, conducted by the National Research Council and led by George Mason University President Alan Merten, himself an information scientist. The findings were reported to Congress in the fall of 2000 and published the following year as *Building a Workforce for the Information Economy.*

The NRC study built upon but went much further than the CRA study. An attempt was made in the NRC study to bring together the analytic perspective of labor economists and other social scientists with the first-hand knowledge of the practitioners—people managing and performing IT work as well as those involved in the education and training of IT workers. After a lengthy analysis of the characteristics of the IT workforce, the report focused on a variety of special topics: older workers, foreign workers, making more effective use of existing workers, longer-term strategies for increasing the IT workforce, and comparative analysis of information technology and biotechnology.

There are many similarities in the findings of the CRA and NRC studies. They both found that there is a tight labor market but insufficient evidence to demonstrate a shortage. They both point to the critical role of the formal educational system in preparing IT workers while at the same time acknowledging that the majority of IT workers do not have formal IT degrees. The two reports share a concern about the seed-corn problem of too many graduate students and faculty leaving for faculty positions, thus weakening the academic system's ability to train the next generation of IT workers.

The Players

While these studies were underway, the national political debate continued—accompanied by intense lobbying to increase the number of foreign workers visas. Who were the players in these debates? As one can imagine, those most in favor of increased numbers of foreign workers were the managers of companies that hired IT workers. The presence of additional foreign workers made it more likely these companies would have access to the talent they needed in their companies, and a larger labor pool helped to control labor costs.

The management perspective was represented in Washington by various groups. National trade associations representing companies employing IT workers, such as ITAA and the Computing Technology Industry Association

(COMPTIA), were the most ardent supporters of increasing foreign workers. Some of the larger companies had their own lobbyists in Washington, while also working through the trade associations. However, the H1-B visa issue tended not to be the highest policy priority for companies whose products or services were outside the IT area—even if the company employed many IT workers. These companies were instead more likely to give policy priority to trade or regulatory issues. In contrast, the IT sector, that is to say those companies that were producing IT products and services, took a strong interest in the visa issue.

Traditional IT companies, such as IBM, had long had their own lobbyists in Washington. They also had representation in Washington through organizations such as the Computer Systems Policy Project, which was an alliance formed in 1989 of the CEOs of a dozen major computer firms to advocate policy positions on trade and technical issues. What was new with this policy debate was the presence of lobbyists representing Silicon Valley. For many years, the business and technical leaders of Silicon Valley had been apolitical. The well-known venture capitalist John Doerr had orchestrated an effort to defeat California Proposition 211, which allowed unlimited financial exposure to shareholder lawsuits. This heady political success led Doerr and others to form TechNet, a political action committee that represented Silicon Valley companies in Washington as well as Sacramento. TechNet became a major lobbying force in Washington, especially during the years of the Clinton presidency; and one of its principal issues was the IT workforce.

If it was management that was in favor of increased foreign workers, one might expect organized labor to be its principal opposition. Traditional labor unions, such as the AFL-CIO, did speak out against the increase in foreign workers. The AFL-CIO had opposed the H-1B visa program since the early 1990s. But high-tech occupations were not highly unionized, so it is not surprising that the traditional labor organizations did not make this issue a high priority. In 1998 the AFL-CIO exacted some changes in the H1-B program to protect American workers in exchange for not opposing the increase in the annual H1-B caps. But these changes in the law had little effect because they were not backed with enforcement mechanisms.

The principal opposition to increased foreign workers came instead from a wide variety of uncoordinated sources, many of them not holding much political clout. Those opposed to increases in H1-B visas included Ralph Nader and Pat Buchanan—strange bedfellows, both with political clout—but for neither of whom was this an issue of primary importance. Others against the increases in the visas included associations of professional engineers, such as the American Electronics Association and the U.S. policy organization of the Institute of Electrical and Electronics Engineers (IEEE-USA), that were interested in protecting jobs for American engineers; individuals such as Norman Matloff, an outspoken computer science professor at University of California-Davis, who believed that his students could not get jobs because software companies were too picky and narrow in the skills they advertised for; individual web sites, such as *www.ZaZona.com/ShameH1B/* run by an Arizona engineer named Rob Sanchez, and *www.familyinjustice.com*, which takes on issues such as domestic violence, family court reform, and the negative impact of foreign temporary visas on U.S. technology workers; the Federation for American Immigration, which was opposed to increases of any kind in immigration numbers; and high-tech worker guilds of minorities under-represented in the IT professions, such as the Coalition for Fair Employment in Silicon Valley, with a sizeable membership of African-American engineers, physicists, and computer scientists.

The positions taken by the computing professional societies on foreign workers is interesting. These professional societies see their most important role as serving individual computer scientists. Mostly they are oriented toward the technical development of the individual practitioner, through the sponsorship of technical conferences, technical publications, and curricular standards. To some degree, each of these societies also takes an interest in the ethics governing the practice of the profession. This role as ethics advocate and omsbudman sometimes placed the professional society in the position of advocating on behalf of the individual engineer in opposition to the employer, or adjudicating a dispute between engineer and employer. Thus one might expect the professional societies to align themselves politically with those protecting jobs for American computer scientists and against an increase in foreign IT workers. However, the situation was complicated by three factors. First, many of these professional organizations, such as the Association for Computing Machinery (ACM) or the IEEE Computer Society, are international organizations. Thus protecting the jobs for American computer scientists at the same time limits opportunities for members of their organizations from other countries. Second, a number of these professional organizations, such as ACM, have industrial as well as individual members; and these industrial members are likely to be in favor of increased foreign visas. Third, many of the members of the professional societies are university faculty members. The demand for IT workers in American industry was attracting significant numbers of faculty members and graduate students to leave academia for industrial positions, mainly at industrial research laboratories or technically oriented start-up companies. Because many computer science and engineering faculty and more than half of the graduate students were foreign-born, there was a concern that limiting visa access would exacerbate problems of finding adequate talent to conduct current academic research projects and training the next generation of IT workers—the latter problem referred to in the literature as "eating the seed corn."

So what position did the computing professional societies take on this issue of foreign workers? They adopted one of three approaches. Some organizations, such as the ACM, chose to remain silent on the issue. A second approach was one taken by CRA, a professional society dominated by its academic members. CRA did not take a position on the visa issue generally, but it strongly advocated exceptions to the H1-B visa regulations that would allow unlimited numbers of visas to individuals working in universities or to individuals holding advanced degrees who wanted to work in industry.

One professional society, the IEEE, spoke up strongly against the increase in temporary foreign workers under the H1-B program—despite the fact that the IEEE was itself an international organization with approximately one-third of its members outside the United States. The policy work of the IEEE is controlled by IEEE USA, which represents the IEEE's US members, not only those who are members of the IEEE's special technical society on computing. IEEE USA was most concerned about protecting unfilled computing jobs for American electrical engineers who were having trouble finding jobs as the defense electronics industry, electric power industry, and consumer electronics industries downsized. The IEEE position evolved to one in which it supported replacement of the H1-B program with a program that provided permanent legal residency for these IT workers. This was more humane in that it reduced the indentured nature of the work, but it was also a position that amounted to the same thing as opposing the H1-B increase in that more green cards was clearly not politically feasible. When there were exceptions carved out in the H1-B program for IT workers working for non-profits, IEEE USA was opposed, wanting all H1-B workers to count towards the cap. The IEEE Computer Society was prohibited by IEEE rules from taking any policy positions. However, many of the individual members of the Computer Society were unhappy with the position taken by IEEE USA, and the Computer Society printed letters to the editor expressing these opposing views in the Society's magazine, *Computer*.

Long-Term Solution

In order to understand the debate of the late 1990s, it is necessary to consider the various alternative solutions considered in response to the political pressure concerning a perceived shortage of IT workers. These fall into three categories of long-, medium-, and short-term solutions.

The long-term solution that was most often considered was to increase the national output of trained workers by enhancing the national educational and training system. This included providing fellowships, scholarships, grants, and loans to existing or prospective students, as well as financial support to educational and training providers in order that they might increase their activities or build new ones.

There were several problems with this solution. The first involved its long-term character since both the politicians and their constituencies preferred a quick fix. Second, there was skepticism about the federal government's ability to forecast and plan for this need accordingly. Having too many trained workers is as much of a problem as having too few, and some people questioned whether the government could really achieve this delicate balance—especially since the supply and demand changed dynamically over time. In the late 1980s, the senior management of NSF had cautioned that there were looming shortfalls of scientists and engineers, using a model that projected a shortage of 675,000 people by 2006. Congress responded quickly by providing NSF with increased funding for science and engineering education programs, and a year later with increased permanent and temporary visas for scientists and engineers. This was one origin of the H1-B visa program (discussed below.) In response to the new funding and call for action from NSF, many universities increased their graduate admissions in these subject areas. Unfortunately, the projection model used by NSF was faulty, and as early as 1992 there was a glut of doctorates in several science and engineering fields. This resulted in congressional hearings in 1992 and harsh criticism of NSF management from several prominent congressional supporters of science and engineering. Both Congress and NSF wanted to avoid a repeat of this embarrassing episode.

The third problem was that universities are organized in ways that make it hard for them to ramp up their degree production in a given field on short notice. A deliberative process for review and decision-making, reluctance of university administrators to make long-term commitments in the form of tenured faculty appointments and new laboratory facilities, the rapidity with which student interests change, and limited discretion in academic budgets all were factors limiting the university's ability to act quickly. The traditional universities are not the only player in the training of IT workers, but they are one of the most important ones. For-profit universities and corporate universities are more able to react quickly to market forces, but much of what they have to offer is in the form of training rather than foundational education.

Despite these three problems, several long-term actions resulted from the political process. Separately from this workforce debate, Congress had requested and President Clinton had formed a Presidential Information Technology Advisory Committee to review the need for further government investment in information technology. The initiative was based largely in the recognition that federal investments in computing research and development of the 1960s and 1970s had fueled the Internet revolution of the 1990s. PITAC recommended a major increase in federal support for computing research across seven federal agencies, but also included in its recommendations the need for support of workforce development. The workforce provision in the PITAC report resulted, for example, in an NSF program to undertake social science research to understand the causes of underrepresentation of women and many minority groups in the IT field, with the recognition that even modest increases in participation of these groups in the IT workforce would go a long ways toward meeting industry's needs for trained workers. While the workforce development provisions of the PITAC recommendations were directly focused on the alleged workforce shortage, the recommendations on increased research funding also had a bearing on the workforce issues. This was because research funding was the most common way to pay for graduate student education. In fact, the research experience itself served as a major component of the graduate student's education.

Another long-term response was to use a portion of the application fees paid by employers who wanted to hire foreign workers on temporary visas through the H1-B program (see below) to provide scholarships to students and funds for educational organizations creating or enhancing IT training programs. Under the H1-B visa program, employers have to pay $1000 (originally $500) for each H1-B visa application. 55 percent of the funds go to the Department of Labor to support worker training programs, 22 percent to the NSF for scholarships, 15 percent to the NSF for K-12 educational programs, and the remaining amount to the Departments of Labor and Justice for processing, enforcement, and complaint management of these programs. Although the bulk of these training funds go to IT, some is spent on training in biotechnology, telecommunications, advanced manufacturing, and health care technologies.

The scholarship program provided funding to schools awarding associate, baccalaureate, and graduate degrees. Awards were made to full-time students with demonstrated academic potential majoring primarily in computer science (38 percent), engineering (37 percent), or mathematics (10 percent). Students received up to $3125 per year for up to four years. By May 2002, NSF had funded 7700 students at 277 schools with approximately $72 million of scholarship funds. While the program is still in its early stages, there is preliminary evidence that the program is attracting students to these majors and that the funds allow many students not to work while attending school, giving them more time for their studies.

Using the H-1B funds, the Department of Labor has provided competitive grants to a variety of training institutions: colleges, for-profit trainers, and companies. The program was organized in a way that required each provider to meet local conditions as determined by a local Workforce Investment Board. By late 2001, 74 training organizations had received $182 million of support from Labor, with an additional $34 million provided in early 2001 to 14 business-related consortia. The training was received by a somewhat diverse population. A study of 16,000 of those trained under this program showed that three-quarters were employed workers upgrading their skills, more than half were between the ages of 22 and 39 years, 69 percent had some previous college education, and 39 percent already held a baccalaureate or higher degree. The trainees included more females (40 percent) and more minorities (e.g., greater than 30 percent African American) than were enrolled in traditional higher education in IT. Significant effort was made to partner with industry in this training, especially with small business; thus there has been active involvement of individual companies, trade associations, and industrial parks in the various training programs.

A number of problems have been identified with the Labor department's training programs. A stated purpose of these programs is to train indigenous Americans to hold the jobs that are typically held by H1-B visa workers. However, many of these programs were training people for lower-level IT occupations. It was estimated that at least 45 percent of the people in these programs would not qualify, upon completion of training, for the jobs typically held by H1-B workers, who are required to have a baccalaureate degree in a relevant technical major. Labor provides only two years of funding to these training programs, with a non-guaranteed possibility of a third year. This is not long enough time to provide an undergraduate education to students who enter without any college credits. Critics of the program are skeptical whether the training provides enough foundational knowledge to enable the graduates of these training programs ever to move up from the lower-level occupations for which the programs prepare them, to higher-level occupations. There is also little coordination among these local training programs, making it more difficult to implement a national training strategy. Nor is their much coordination of the Labor department's training programs with the NSF scholarship program, or with programs for IT employers and employees offered by the Department of Commerce (funded from sources other than H1-B application fees). Many of the training programs are having trouble moving past the start-up phase. With the downturn in the economy and the dot-com crash in the past several years, companies have been reluctant to continue their participation in these training programs, seeing no need for them in the short term.

Medium-Term Solution

The proposed medium-term political solution to the workforce problem was training tax credits. Employers prefer to hire workers with exactly the right skill set to carry out the company's current projects, so that the workers can "hit the ground running." This has seemed to be particularly important to start-up firms, where there is often only a small time window of opportunity to enter a market. Unfortunately, there is often a shortage of available workers with exactly the right skill mix. Larger, more established companies in the high-tech area can invest in training for new and prospective workers; but smaller companies claim they cannot afford either the time or the cost. Employers have also expressed concern that any investment they make in training their workers may turn out to be counterproductive when the trained employee becomes more marketable and quits to join a competing firm. However, training is more palatable to employers if the cost is off-set by tax credits. In a similar vein, there was a political interest in offering tax credits to individuals to invest in training that would make them more qualified workers.

Considerable political effort was made to obtain training tax credits at both the state and federal levels. The lead advocate has been the Technology Workforce Coalition, representing a dozen IT trade associations and more than 500 companies, orchestrated by CompTIA) Through the efforts of CompTIA and others, The Technology Training Tax Credit Act became law in Arizona, providing 100 percent tax credits up to $1500 per year per person towards the cost of training for IT skills. Maryland also passed a similar bill, Critical Skills and Occupations—Income Tax Credits. Similar legislation (The Technology Education and Training Act of 2001) was introduced at the federal level, but perhaps for several reasons it never was passed into law. Tax revision costs too much political capital at the federal level for something that has not been a top priority for the Republican administration. Moreover, by 2001, the dot-com crash had lessened the demand for IT workers significantly.

Short-Term Solution

The greatest attention has been given to finding a short-term political solution to the worker issue, in particular by increasing the number of temporary foreign workers. This was also, by far, the most contentious solution proposed, with intensive lobbying from a variety of viewpoints. One attraction of this solution was that it could be carried out through the simple act of increasing the annual quota of foreign workers who could come to the United States under the existing H1-B temporary visa program, and by making some other minor modifications to that program.

The Immigration Act of 1990 was a major piece of immigration legislation that swung the policy pendulum slightly toward the immigration of skilled workers, away from the more traditional goal of reuniting families. Employment-based immigrants rose from 58,000 to 140,000 per year. The most important aspect of the bill with respect to the IT worker debate was the H1-B visa program for temporary nonimmigrant workers in specialty occupations. By the late 1990s, the majority of those visas were being awarded to people who worked in high-tech and especially computer-related occupations, as well as to accountants, architects, and surveyors. In the first few years of the program, many of the visas went instead to foreign specialty cooks, physical therapists, and fashion models. There had been a nursing shortage in the United States in the late 1980s, and the H1-A visa program for temporary, nonimmigrant (that is, non-permanent) registered nurses was enacted in 1989 as a response. The H1-B visa program was structured after the H1-A program, without the benefit of time to see how successful the H1-A program turned out to be. With hindsight, it is possible to see that the H1-A program was at best a moderate success. It made only a slight dent in the shortage of nurses. Most of the H1-A visa holders ended up in major U.S. cities, especially New York, rather than spread out across the country. The presence of these nurses lessened the pressure on employers to find long-term solutions to their nursing shortages and these shortages flared up again during the following decade (see Table 10–1).

Table 10–1   Visa Designation and Descriptions

Type of Nonimmigrant Visa	Description of Visa
E	Treaty Traders and Investors
F	Foreign Students
H	Specialty Occupation
J	Exchange Visitors and Spouses
L	Intra-company Transferees
TN	NAFTA Visas

Under the H1-B program, a specialty worker could work in the United States for two successive three-year terms before being required to leave the United States. During this period of time, the worker could—and many did—apply for permanent residence (green card) status, under which they could remain and work indefinitely in the United States as a non-citizen. The H1-B visa ties a worker to a specific employer. It is possible for an H1-B worker to transfer to another employer, but to do this had the disadvantage that any application for permanent residence status had to be restarted. The process of obtaining a green card is lengthy, seldom taking less than three years, and often taking much longer. Changing employers while here on an H1-B visa often meant that the visa holder was forced to leave the United States before the green card application process was completed. This was especially true for workers from the People's Republic of China and India because annual per-country limits on green cards typically added two to four years to the process for citizens of those countries. Given this strong disincentive, the vast majority of H1-B workers remained with their initial U.S. employer, no matter how unsatisfactory the arrangement.

Under intense political pressure, Congress twice modified the conditions on the H1-B visa program. As originally enacted, the law capped the annual number of H1-B visas at 65,000. The American Competitiveness and Workforce Improvement Act of 1998 temporarily raised the cap on H1-B visas to 115,000 for FY 1999 and FY 2000, dropping to 107,500 in FY 2001, and returning to the original amount of 65,000 in FY 2002. However, even with the increased quotas, the caps were hit long before the end of the fiscal year, meaning that employers had to sit idly by waiting for the new fiscal year to begin before progress could be made on their application. Not surprisingly, there was renewed pressure to increase the limits once again. This resulted in the Competitiveness in the Twenty-First Century Act of 2000, which further raised the caps.

Aspects of the H1-B program were under the jurisdiction of three federal agencies, creating a variety of coordination and control problems. An employer of a potential H1-B visa worker had to file a labor condition application that indicated that the foreign worker would be paid prevailing wages for this specific occupation and specific geographic region, the working conditions of U.S. workers would not be adversely affected (e.g., loss of jobs or reduced pay scale) from the hire of the H1-B worker, and there was no strike or lockout underway on the employer's premises. However, Labor did not have adequate staff to enforce these rules, and there were no provisions (such as stiff fines) to punish employers who did not comply. The employer would send in its labor condition application, as certified by Labor, to the Immigration and Naturalization Service (INS), together with the name of the specific individual it wished to hire, the candidate's educational and work credentials, and an application fee. Once the application was approved by INS, if the prospective worker was not already in the United States (for example, on a student visa), then the applicant must apply for a travel visa to enter the United States, which required approval from the Department of State. INS was responsible for tracking the number and status of H1-B applications, including assurance that the annual caps set by Congress were not being exceeded. Not only was INS unable to provide reliable data to Congress about the types of people being admitted under the H1-B program, it had trouble keeping track of the number of H1-B visas awarded each year—one year awarding too many visas and another year cutting off the applications before the quota was met.

Definitions and Data

Virtually everyone involved in the policy debates over the IT workforce in the last few years has decried the ambiguity over who is an IT worker. IT work encompasses many different occupations, requiring a wide array of skill sets. The skills required to be a successful help-desk operator, for example, are very different from those needed by a creator of new information technology. Is everyone who works for an IT company, including the person who cuts the lawn at IBM, to be considered an IT worker? Should someone who uses devices that have IT embedded in them, such as an automobile mechanic using diagnostic tools, be considered an IT worker? The CRA and NRC studies mentioned above arrived at somewhat similar definitions, which are increasingly those used in recent policy discussions about the IT workforce. (See Figure 10–1)

The CRA definition is the more intuitive one. Each IT-related occupation can be placed on a two-dimensional grid, depending on the amount of IT knowledge and the amount of business/industry knowledge required to perform the work in that occupation. Those occupations requiring more business knowledge than IT knowledge are considered IT-enabled jobs rather than IT jobs and are dismissed from the

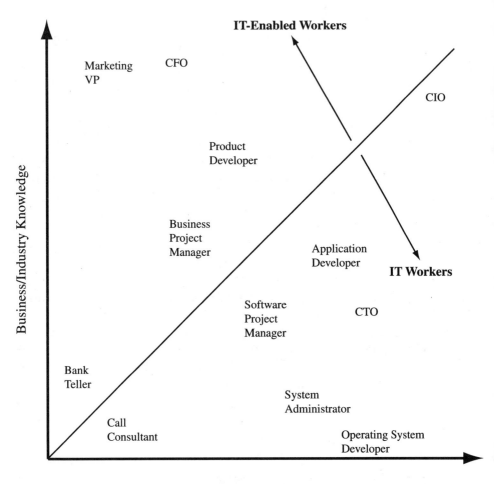

Figure 10–1   IT workers v. IT-enabled workers.

policy considerations of shortages of IT workers. Those occupations that require more IT knowledge than business knowledge to perform the job are considered IT work—making this the class of occupations under primary consideration in policy debates about IT worker shortages. (More and more occupations are requiring some IT skills, and training for the population for these IT-enabled jobs is a different political issue that has yet to be confronted.) Under the CRA scheme, the car mechanic would be an IT-enabled worker but not an IT worker, while the IBM groundskeeper would be neither an IT worker nor an IT-enabled worker.

The NRC study provides a definition that may be slightly less intuitive than the one from CRA, but which has the merit of mapping reasonably well on to the occupational categories of the Bureau of Labor Statistics—the best source of long-term data on the IT workforce. The NRC study identifies two categories of

IT workers. Category 1 work "involves the development, creation, specification, design, and testing of an IT good or service, or the development of system-wide applications or services." Category 1 work also includes IT research. What the committee calls Category 2 work primarily involves the "application, adaptation, configuration, support, or implementation of IT products or services designed or developed by others....the term 'IT worker' without qualification refers to both Category 1 and Category 2 workers." (Executive Summary, Building a Workforce for the Information Economy). This definition of an IT worker is roughly congruent to the one given in the CRA study.

The policy discussion has been hampered by the lack of good data as well as by good definitions. The federal data is far superior to most private data. The federal data is objectively gathered and based on large samples or entire population surveys using well-established methods and procedures. Nonetheless, the federal data had some shortcomings.

There is a fundamental tension between keeping the occupational categories up to date and being able to provide consistent longitudinal data. Bureau of Labor Statistics occupational categories were not updated between 1980 and 1998, thus leading to years of data collection in which jobs associated with multimedia and the Internet, for example, were hard to capture.

The federal data often takes two or three years to appear and thus is often not timely for making decisions about the fast-paced high-tech industry.

- Some of the occupational categories are too broad and ill-defined to use for understanding the policy issues well. For example, all programmers are classified together in the federal data, but this does not mean that all programmers have similar enough skills that they could hold any programming job. It has been demonstrated that many programmers trained in old programming languages such as COBOL were not able to learn to program in modern languages such as Java because of the differences in programming methodologies.

- The levels of data aggregation have often been too broad, such as the various federal data in which computer science is sometimes grouped together with mathematics, or engineering, or the physical sciences—fields that may have different policy issues from those of computer scientists.

- Data was often gathered in relation to earlier issues of policy concern that does not always serve well for current policy decisions. There is not data, for example, on certificate programs or corporate universities, which have become major supply paths for IT workers in recent years, but there is good data on B.A., M.A., and Ph.D. production, which have been important steps in the traditional career paths.

- Even when the policy issue has been present for some years, the data was sometimes inadequate. This was the case with data about the number of H1-B visa holders, in what occupations, from which countries, and for what kinds of companies. Demand data is particularly weak.

- Finally, there is a mismatch between the categories reported in supply and demand data as reported by different federal agencies, making it difficult to compare the two.

Was There a Shortage?

The policy debate has also been hampered by a clear meaning of the term "shortage" and lack of lack of good methods for determining whether there is a mismatch between supply and demand. Workforce shortage, which appears at first to be a very simple concept, turns out to be devilishly complicated. One would assume that the policy goal is to achieve a relatively close match between the number of qualified people who are interested in work and the number of available positions. But what does it mean to be "qualified"? Does it mean having a degree in computer science or some other computing-related discipline? What type of degree? Does it mean having work experience? Does it mean having already worked with the specific set of technologies the person will use in the new job? A worker who has the skill set to qualify for one IT job might well not be qualified for a different IT job. Similarly, what does it mean to be "interested in work"? Workers might be interested in obtaining an IT job, but they might be interested only if the job is paid at a certain salary level, stock options are offered, the work is located in a certain geographic region, the employer is of a certain type, or the work has a certain technological challenge or cachet.

Specifying demand is just as difficult as specifying supply. Does one count vacancies? Perhaps not, because it is a common practice for management to have permanently open lines (so-called "evergreen positions") so that they might act quickly to hire if an individual with exceptional credentials comes along, even though they otherwise had no intention of filling the position. Does one count job advertisements? Perhaps not, since they can either over- or under-represent demand. There might be an advertisement for an evergreen position, or an ad for a single job might be placed in multiple publications—thus leading to an over-count of demand. On the other hand, a single ad might be placed in a newspaper when the company in fact is looking to hire more than one person—thus leading to an under count of demand.

From these brief comments, it is clear that it is difficult to make a simple count of people available and jobs available to see if there is a match between supply and demand. Partly because of these difficulties, labor economists use a set of metrics that are indicators of secondary effects commonly occurring when there is an imbalance between worker supply and demand. These include unemployment rates from federal statistics, permanent or temporary labor certificates issued by the Department of Labor, wage growth (including salary, stock options, signing bonuses, and benefits), worker demand projections from the Bureau of Labor Statistics, churn rate (the number of people leaving one job for another in the same profession), and overtime hours reported.

Based on a consideration of these metrics, both the CRA and the NRC studies stopped short of proclaiming a shortage of IT workers. Both reports instead claimed that there were clear signs through these secondary economic metrics of a tightness in the IT labor market, but insufficient evidence of more jobs than people to fill them. These conclusions, however, did not imply that no political action was justified, nor were they intended to do so. The political pressure exerted by employers was sufficiently great that it is not at all surprising that political action was taken and the H1-B visa quotas raised.

Changing Environment

Since the two legislative increases in the H1-B visas quotas, the environment for IT work has changed significantly. The dot-com crash has meant that many of the start-up companies that were driving the Internet revolution have gone out of business. Even well-established high-tech companies such as Sun Microsystems have lost a large fraction of their stock market valuation and have had to scale back on their employment. Seldom today do we hear plaintive remarks from company managers that they cannot find enough high-quality workers. Indeed, the pendulum has swung to the point that many IT workers are having trouble finding jobs that they deem desirable. More commonly we are hearing complaints from U.S. citizens about foreign workers who are filling jobs that might have been theirs.

Statistics prepared on behalf of the ITAA in September 2002 give a snapshot of the current national labor market for IT (Quarterly update to the ITAA annual labor report, Bouncing Back, May 2002). Despite the recession and the dot-com crash, the number of IT workers in the United States had grown since January 2002 by 85,000—to almost 10 million (9.98 million). As always, there were dismissals (697,000), but these were exceeded by hires (782,000). In both January 2002 and July 2002, the employers of IT workers had expected the IT workforce to grow in the following 12 months, but the slow recovery of the economy led to lower expectations by July. Indeed, in July employers estimated the growth at 27 percent lower than they had in January (1.148 versus 834,000 position growth). From January to December 2001, dismissals had exceeded hirings (2.6 million to 2.1 million). However, if one looks at the situation only six months later, for the overlapping period of July 2001 through June 2002, the situation had reversed and hires (1.6 million) surpassed dismissals (1.4 million). Although hirings were off by 25 percent in this later 12-month period (July 2001–June 2002), dismissals were down by 48 percent. This suggests that most job-cutting was over.

The numbers given in the preceding paragraph were for all IT workers, independent of the market sector in which they worked. If one looks only at the IT sector (that is, at companies that produce IT products and services), the situation is much bleaker. In January 2002, the IT sector was hiring 20 percent of all IT workers in the United States. However, by July 2002, the IT sector was only hiring 5 percent of these workers. Hirings during this period were off by 85 percent in large IT companies and by 79 percent in small IT companies. Because of these drop-offs in hiring in the IT sector, the number of IT workers working within the IT sector had dropped to only 9 percent. The other 91 percent of the IT workers in the United States were working in government, nonprofit, and industries other than the IT industries. In the three months prior to this ITAA report, almost two-thirds of the IT hires were attributable to three of the many dozens of IT occupations: technical support specialists (147,000), web developers (93,000), and network designers/administrators (47,000).

This data suggests there is some hope for a turnaround in the short-term for IT work overall, if not in the IT sector itself. Should the nation come out of the recession, as economists have been predicting for some months, there will likely be continued growth in the employment of IT workers in the United States. The threats of war and terrorism seem to be the most salient problems thwarting economic recovery.

The long-term picture continues to look rosy. Technological and economic development have been driven since the 1960s by Moore's Law, the empirical law that indicates that the number of switches that can be placed on a unit area of a chip doubles every 18 months. There are good technological reasons to believe that Moore's Law will continue in force for at least another decade, bringing with it regular increases in price-performance, product innovation, and economic growth. We seem to be only at the beginning of the Internet era. Even the personal computer, which has now become a commodity product, seems to have vast room for improvement. It is not nearly as user-friendly as other complicated consumer goods, such as the television or the automobile; we have not yet achieved the marketing mantra of "plug and play". And there continue to be a steady stream of new scientific and technological advances such as bioinformatics, quantum computing, and cluster supercomputing.

What is the meaning of all this for IT policy related to the workforce? Although the IT field has had a half century of rapid, perhaps unprecedented growth, and there are good reasons to expect this growth to continue, even the computer field has had periods of economic downturn. In these periods, there is likely to be a reduced demand for labor. The fluctuations in demand are likely to change more rapidly than the worker supply system can react to them, for it takes some time to recognize the demand and more time to train workers with the right set of skills. Temporary foreign workers thus seem to be a good addition to the marketplace, to provide rapid response to spikes in demand. However, temporary workers fit better as a temporary solution, not a permanent fixture in the marketplace. The United States would be better off to develop a permanent workforce that, as best as it can foretell, matches the present and near future demand.

Policy Implications

The remarks in the last section have a number of policy implications. There is a sunset clause in the current legislation that increased the annual cap on the number of H1-B workers. Given the continued weak economy and the particular weakness in the IT sector, there seems no reason—and no strong political will—at this time to pass legislation to continue the elevated caps. This is not to say that the H1-B caps should not be elevated again at some time in the future when there is increased demand.

Some companies, especially some programming companies headquartered in India, found a loophole in the immigration rules when the job market was hot. They used L visas, which are intended to bring intra-company transfers for multinational companies to the United States, as a way to supplement their ranks of temporary workers beyond those available through the H1-B program. There is some sentiment today to try to close this loophole, for example either by restricting the uses of L visas in some way or at least placing a cap on the number of L visas that can be awarded each year.

There is value in having a small number of temporary nonimmigrant workers enter the United States each year. The United States has been built on incorporating foreign workers into its permanent workforce, and there seems to be no reason to change this winning strategy. It is a good opportunity for workers from other countries to see if the United States is the right place for them. It is a chance for U.S. employers to try out workers and see if they would like to have these workers on their payroll permanently. It is also a way to acquire specific technical skills that may be lacking in the U.S. workforce. Even when there was not generally a shortage of IT workers, there was often a shortage of workers with very specific skills, which could be met by hiring foreign workers.

The federal government needs to do a better job at tracking present and future supply and demand for IT workers. This involves improving the speed at which data is made available, having analysts who continually monitor the situation and report to Congress and relevant agencies, coordinating the data collection better between federal agencies (for example, so that the demand data from one agency can be compared easily with the supply data from another agency), and updating the job categories in a way that reflects innovation in information technology but also still allows longitudinal analysis.

Assuming that over the long run there is sustained growth in the demand for IT workers, the United States needs to do a better job of training its own workers. This means finding ways to interest more U.S. students in the subject and keep them in the appropriate educational track that will prepare them for this work. It also means finding ways to increase the participation of under-represented portions of the U.S. population in IT careers—especially women and African Americans, Hispanic Americans, and Native Americans.

There also needs to be further study of the increased reliance on short courses and certificate programs as a way to an IT career. A number of the regional economic development programs that gave unemployed citizens their initial training in IT have not succeeded very well in getting employers to hire their graduates, presumably because they do not know enough IT or do not have other important work skills such as communication, teamwork, and analytic skills. The higher education community is skeptical of the value of short courses and certificates; they believe that students may learn some very specific skills but not the foundational knowledge needed to keep pace in a rapidly changing field.

There has been much discussion of the problem of age discrimination in the IT field. The NRC study addressed this problem carefully. It noted that IT workers over the age of 40 were more likely to lose their jobs than younger IT workers. However, it also found that older IT workers are just as likely to find new IT jobs as younger workers, and that it takes individuals in the two groups about the same amount of time to find a new IT job. The NRC committee noted the paucity of data. Clearly there is policy need to watch this issue and collect more data.

Policy makers need to pay attention to the growing problem of IT jobs being moved overseas. There is presently lots of discussion but little hard data about the issue. Attention also needs to be paid to retaining organizational skills in the full array of IT work at the firm, regional, and national levels. Silicon Valley continues to be outstanding at cutting-edge technology, but it is less well positioned to lead in the production and sales of mature IT technology. When the microcomputer became a commodity product in the 1990s, this industry moved out of Silicon Valley to other regions of the United States (e.g., Austin, Texas) and overseas (especially hardware manufacture in Taiwan). We are seeing increasing amounts of programming and software testing and maintenance moved to Asia.

To a great degree, recent policy on IT generally and on IT workers in particular has been shaped by the concerns of Silicon Valley firms. There is some justification for the influence of Silicon Valley on the policy agenda, given what an important driver Silicon Valley has been of the national economy. However, when it is remembered that less than 10 percent of the IT workforce works in the IT sector, it may seem appropriate to listen to a wider community of participants and adjust the policy concerns to reflect the issues of greatest concern to those employers of the other 90 percent of IT workers as well.

In the 1990s, the federal government was faced with a serious IT worker problem. Its cadre of programmers were aging and fast approaching retirement age, much of the programming for government systems was on legacy computer systems using older programming languages, such as Fortran and Cobol that were not of interest to young IT workers, and salaries and work conditions were not competitive with those in industry. Federal agencies have done a lot in the past several years to improve salaries and working conditions and actively recruit IT workers; and government service generally seems more attractive after the 9-11 terrorist attacks. So, the federal government is beginning to gain control over its workforce for its traditional IT work. However, there is a new set of workforce issues arising with homeland defense. This country has an acute shortage of working professionals and of education and training programs in the areas of computer security and information assurance. Programs have been recently funded by the National Security Agency and administered by the NSF to provide scholarships for computer security students, as well as to help colleges and universities build up their curricula in this area. But this will continue to be a major workforce challenge for the next few years.

Further Reading

Peter Freeman and William Aspray, *The Supply of Information Technology Workers in the United States*. Washington, D.C.: Computing Research Association, 1999.

National Research Council, Committee on Workforce Needs in Information Technology. *Building a Workforce for the Information Economy*. Washington, D.C.: National Academy Press, 2001.

Information Technology Association of America, various studies and reports on the IT workforce, *www.itaa.org*

U.S. Department of Commerce, Office of Technology Policy, *America's New Deficit: The Shortage of Information Technology Workers*, 1997.

Acronyms

AAAS	American Association for the Advancement of Science
AAU	Association of American Universities
ACLU	American Civil Liberties Union
ACM	Association for Computing Machinery
AEC	Atomic Energy Commission
AEI	American Enterprise Institute
AFL-CIO	American Federation of Labor-Congress of Industrial Organizations
AOL	America Online
ARPA	Advanced Research Projects Agency (later DARPA)
BA	Bachelor of Arts
BLS	Bureau of Labor Statistics
CalTech	California Institute of Technology
CAWMSET	Commission on the Advancement of Women and Minorities in Science, Engineering and Technology
CBO	Congressional Budget Office
CD	Compact Disc
CDA	Communications Decency Act
CEA	Consumer Electronics Association

CEO	Chief Executive Officer
CER	Coordinated Experimental Research Program
CFAA	Computer Fraud and Abuse Act
CIAO	Critical Infrastructure Assurance Office
CIPA	Children's Internet Protection Act
CISE	Computing Information Science and Engineering Directorate
COBOL	Common Business Oriented Language
COMPTIA	Computing Technology Industry Association
COPA	Child Online Protection Act
COPPA	Children Online Privacy Protection Act
CPC	Congressional Privacy Caucus
CPPA	Child Pornography Prevention Act
CPST	Commission on Professionals in Science and Technology
CRA	Computing Research Association
CRADA	Cooperative Research and Development Agreements
CSPP	Computer Systems Policy Project
CSS	Content Scramble System
CSSPAB	Computer System Security and Privacy Advisory Board
CTC	Community Technology Center
CTEA	Copyright Term Extension Act
DARPA	Defense Advanced Research Projects Agency
DAT	Digital Audio Tape
DCS	Distribution Control Centers
DHS	Department of Homeland Security
DJ	Disk Jockey
DMCA	Digital Millennium Copyright Act
DOD	Department of Defense
DRM	Digital Rights Management
DVD	Digital Versatile Disc

ECPA	Electronic Communication Privacy Act
ED	Department of Education
EFF	Electronic Frontier Foundation
ENIAC	Electronic Numerical Integrator and Computer
EPIC	Electronic Privacy Information Center
ESA	Economics and Statistics Administration
ESEA	Elementary and Secondary Education Act
EU	European Union
FEC	Federal Elections Commission
FISMA	Federal Information Security Management Act of 2002
FOIA	Freedom of Information Act
FTC	Federal Trade Commission
FutureMAP	Future Market Applied to Prediction
GAISP	Generally Accepted Information Security Principles
GDP	Gross Domestic Product
GE	General Electric
GISRA	Government Information Security Reform Act of 2000
GLBA	Gramm-Leach-Bliley Financial Services Modernization Act
H1-A	U.S. Nursing Visa Program
H1-B	U.S. Specialty Worker Visa Program
HEW	Department of Health, Education, and Welfare
HHS	Department of Health and Human Services
HIPAA	Health Insurance Portability and Accountability Act
HPCCI	High Performance Computing and Communications Initiative
HPP	Health Privacy Project
HHS	Department of Health and Human Services
HTML	Hypertext Markup Language
HTTP	Hypertext Transfer Protocol
HUD	Department of Housing and Urban Development

IBM	International Business Machines
IEEE	Institute of Electrical and Electronics Engineers, Inc.
IITF	Information Infrastructure Task Force
INS	Immigration and Naturalization Service
IPTO	Information Processing Techniques Office
ISAC	Information Sharing and Analysis Center
ISP	Internet Service Provider
ISSA	Information Systems Security Association
IT	Information Technology
ITAA	Information Technology Association of America
K–12	Kindergarten through Twelfth Grade (public education)
LEDX	Law Enforcement Data Exchange
LICRA	International League Against Racism and Anti-Semitism
LOGIC	Local Government Information Control
MA	Master of Arts
MIT	Massachusetts Institute of Technology
MOSIS	Metal Oxide Silicon Implementation Service
MPAA	Motion Picture Association of America
MRAP	French Movement Against Racism
NAACP	National Association for the Advancement of Colored People
NASA	National Aeronautics and Space Administration
NBS	National Bureau of Standards—later NIST
NCES	National Center for Education Statistics
NCLB	No Child Left Behind Act
NCMEC	National Center for Missing & Exploited Children
NCR	National Cash Register
NIH	National Institutes of Health
NII	National Information Infrastructure
NIPC	National Infrastructure Protection Center

NIST	National Institute of Standards and Technology
NITRD	Networking and Information Technology Research and Development
NOAA	National Oceanic and Atmospheric Administration
NRC	National Research Council
NSA	National Security Agency
NSF	National Science Foundation
NTIA	National Telecommunications and Information Administration
OMB	Office of Management and Budget
ONR	Office of Naval Research
OPG	Online Policy Group
P2P	Peer-to-Peer
P3P	Platform for Privacy Preference
PARC	Xerox Palo Alto Research Center
PCAST	President's Council of Advisors for Science and Technology
PCCIP	Presidential Commission on Critical Infrastructure Protection
PCIS	Partnership for Critical Infrastructure Security
PDD	Presidential Decision Directive
PETs	Privacy-Enhancing Technologies
Ph.D.	Doctor of Philosophy
PII	Personally-Identifiable Information
PIN	Personal Identification Number
PITAC	Presidential Information Technology Advisory Committee
PITs	Privacy-Invading Technologies
POM	Policy Analysis Market
PRC	Privacy Rights Clearinghouse
PROTECT	Prosecutorial Remedies and Other Tools to End The Exploitation of Children Act
RFPA	Right to Financial Privacy Act

RIAA	Recording Industry Association of America
SAUC	Students Against University Censorship
SCADA	Supervisory Control and Data Acquisition System
SCI	Strategic Computing Initiative
SCMS	Serial Copy Management System
SDMI	Secure Digital Music Initiative
SERVE	Secure Electronic Registration and Voting Experiment
SIA	Semiconductor Industry Association
SIAM	Society of Industrial and Applied Mathematics
SSRC	Social Science Research Council
TIA	Terrorist (formerly Total) Information Awareness
TOP	Technology Opportunity Program
U.S. PIRG	U.S. Public Interest Research Group
UCITA	Uniform Computer Information Transactions Act
UEJF	Union of French Jewish Students
USA PATRIOT	Uniting and Strengthening America by Providing Appropriate Tools Required to Intercept and Obstruct Terrorism
USSR	Union of Soviet Socialist Republics
VA	Veterans' Administration
VCR	Video Cassette Recorder
VLSI	Very Large Scale Integrated Circuits
W3C	World Wide Web Consortium
WIPO	World Intellectual Property Organization
WTO	World Trade Organization
Y2K	Year 2000 (software crisis)

Index

A